McGraw-Hill
Mathematics

Daily Homework Practice

3

McGraw-Hill
School Division

New York   Farmington

McGraw-Hill School Division 🐝

*A Division of The **McGraw·Hill** Companies*

Copyright © McGraw-Hill School Division,
a Division of the Educational and Professional Publishing Group of The McGraw-Hill Companies, Inc.
All rights reserved.

McGraw-Hill School Division
Two Penn Plaza
New York, New York 10121-2298

Printed in the United States of America

ISBN 0-02-100285-1 / 2

1 2 3 4 5 6 7 8 9 024 05 04 03 02 01 00

GRADE 3
Contents

Chapter 8: Division Facts

Chapter 9: Multiply by 1-Digit Numbers

Chapter 10: Divide by 1-Digit Numbers

Chapter 11: Measurement

Chapter 12: Geometry

Chapter 13: Fractions and Probability

Chapter 14: Decimals

Name_____

 ▶ **Explore Place Value**

Write the number shown in each place-value chart.

1.

hundreds	tens	ones
7	2	6

_____

2.

hundreds	tens	ones
5	0	9

_____

Use a place-value chart to show each number.

3. 382

hundreds	tens	ones

4. 91

hundreds	tens	ones

Write each number. Use a place-value chart to help.

5. 6 hundreds, 2 tens, 2 ones _____

6. 1 hundred, 5 tens _____

7. 9 hundreds, 1 ten, 4 ones _____

Solve.

8. In the year 2027, the United States will be 251 years old. How would you write this number in a place-value chart?

_____

Spiral Review

9. 6 + 3 = _____   **10.** 8 + 5 = _____   **11.** 9 + 8 = _____

Name _____

 **I·2** **Place Value Through Thousands**

Write each number in standard form.

1.

thousands	hundreds	tens	ones
2	0	8	7

_____

2. 9 hundreds, 2 tens, 6 ones _____

3. 4 + 700 + 6,000 + 70 _____

Write the word name for each number.

4. 623 _____

5. 3,485 _____

Write each number in expanded form.

6. 381 _____   **7.** 5,879 _____

Problem Solving

8. A used-car dealer has a station wagon for sale for $7,399. How can you write that number in expanded form? As a word name?

_____

Spiral Review

9. 7 − 4 = _____   **10.** 15 − 7 = _____

Name_____

 **1·3** **Place Value Through Hundred Thousands**

Write the value of each underlined digit.

1. 7<u>4</u> _____

2. <u>6</u>83 _____

3. 40,<u>1</u>45 _____

4. <u>3</u>92,440 _____

Write the value of 3 in each number.

5. 37 _____

6. 603 _____

7. 43,928 _____

8. 721,317 _____

Write the digit in each place named.

9. 1,428 (tens) _____

10. 67,300 (thousands) _____

Problem Solving

11. The Two-Wheeler Bike Company has sold about 125,600 bikes over the past ten years. How can you write this number in expanded form?

_____

12. In 1990, there were 453,588 people living in the State of Wyoming. How can you show the number of people in expanded form?

_____

Spiral Review

Skip count to find each missing number. Tell how you skip counted.

13. 6, _____, 12, 15, _____   _____

14. 47, 57, _____, _____, 87   _____

Name_____

 **I·4** **Compare and Order Whole Numbers**

Compare. Write >, <, or =.

1. 77 _____ 88   **2.** 37 _____ 73   **3.** 112 _____ 89

4. 65 _____ 156   **5.** 233 _____ 54   **6.** 132 _____ 132

Order from least to greatest.

7. 4,841; 4,739; 5,160 _____

8. 3,288; 3,324; 3,210 _____

Order from greatest to least.

9. 6,392; 5,821; 6,184 _____

10. 5,497; 5,509; 5,675 _____

Problem Solving

11. Wes collected 217 cans for recycling, Mariana collected 271, and Scott collected 224. Order the three names from most to least cans collected.

_____

12. In the 1968 election, 191 electoral votes went to Hubert Humphrey, 46 to George Wallace, and 301 to Richard Nixon. Who won the election?

_____

Spiral Review

Write each number in standard form.

13. 8 hundreds, 2 tens, 9 ones _____   **14.** 7 hundreds, 7 ones _____

Name_____

1·5 **Problem Solving: Reading for Math**
Using the Four-Step Process

Solve. Use the four-step process.

1. Jordan has 20 snowballs. He makes 10 more. Then he makes 10 more again. How many snowballs does he now have?

   _____

2. From 3 days of yard work, Kendra earns $40. In the next two days, she earns $20 each day. How much money does she earn over 5 days?

   _____

3. In a 3-minute bicycle race, Tomás rides 650 meters. Joan rides 100 meters more than Tomás. Drew rides 250 meters less than Joan. How far does each person ride? Who came in last?

   _____

Use data from the table for problems 4–5.

4. Which village has the most people?

   _____

Village	Number of People
Hillside	670
Sun River	920
Millbrook	880

5. Suppose that 100 people move to Hillside. How many people would Hillside have now?

   _____

Spiral Review

Write each number in expanded form.

6. 739 _____   7. 6,411 _____

Name_____

 **1·6** **Round to Tens, Hundreds, and Thousands**

Round to the nearest ten or ten dollars.

1. 42 _____

2. $85 _____

3. $793 _____

4. 1,776 _____

5. $4,202 _____

6. 7,474 _____

Round to the nearest hundred or hundred dollars.

7. 108 _____

8. $577 _____

9. 2,366 _____

10. $3,929 _____

11. 16,437 _____

12. $29,798 _____

Round to the nearest thousand or thousand dollars.

13. 3,140 _____

14. $4,890 _____

15. $6,507 _____

16. 11,414 _____

17. 49,225 _____

18. $64,477 _____

Problem Solving

19. Skis are on sale for $149 a pair. What is this price rounded to the nearest ten? _____

20. A CD player is reduced to $298. What is this price rounded to the nearest hundred? _____

Spiral Review

21. 18 + 61 = _____ **22.** 25 + 38 = _____ **23.** 47 + 33 = _____

1·7 Problem Solving: Strategy
Make a Table

On a separate sheet of paper, make a table to solve problems 1–2.

Meal Choices at Lunchtime

Wednesday	Thursday	Friday
Hamburger	Sloppy Joe	Pizza
Chicken Salad	Tuna Casserole	Meatloaf
Eggplant	Green Salad	Green Salad
Grilled Cheese	Peanut Butter Sandwich	
	Grilled Cheese	

1. Which day had the fewest choices? _____

2. How many more dishes would you have to serve on Wednesday for it

   to be the day with the most meal choices? _____

Mixed Strategy Review

Use data from the table for problems 3–4.

3. If each student chose only one activity, how many students answered the survey?

   _____

Usual After-School Activities

Activity	Tally	Number
Playing Outside	ⅢⅢ ⅢⅢ Ⅰ	11
Reading	ⅢⅢ Ⅰ	6
Watching Television	ⅢⅢ ⅢⅢ	8

4. How many fewer students participate in the most popular activity than in the other two activities combined?

   _____

Spiral Review

5. $39 - 14 =$ _____

6. $74 - 55 =$ _____

7. $62 - 28 =$ _____

8. $57 - 18 =$ _____

Name_____

 **1·8** **Explore Money**

Complete the table. Use play money to help.

	Item	Cost	You Give	Change
1.	Yogurt	$0.53	$1.00	_____
2.	Sunflower Seeds	$0.87	$1.00	_____
3.	Orange Juice	$2.65	$3.00	_____
4.	Bread	$1.49	$5.00	_____
5.	Cheese	$3.22	$4.00	_____
6.	Raisins	$1.02	$5.00	_____
7.	Apple	$0.33	$1.00	_____
8.	Milk	$0.61	$5.00	_____

Solve. Use play money to help.

9. A package of pretzels costs $0.73. If you pay with a dollar bill, how

   much change should you get back? _____

10. At the video store, you give the clerk $5.00 for a rental that costs

    $3.83. How much change should you get back? _____

Spiral Review

Write the number that is 10 less than each number shown.

11. 70 _____   12. 33 _____   13. 185 _____   14. 313 _____

Name_____

 **1·9** ▶ **Count Money and Make Change**

Write the money amount.

1.

2.

_____   _____

Find the amount of change. List the coins and bills you could get.

3. Cost: $1.65. You Give: $5.

4. Cost: $3.94. You Give: $10.

_____   _____

_____   _____

Problem Solving

5. Gabriela bought a map of the city for $3.60. She paid with a $5 bill and was given a dollar and two quarters in change. Is this the correct? If not, how is it wrong?

_____

Spiral Review

Compare. Write >, <, or =.

6. 95 _____ 59   7. 122 _____ 212   8. 234 _____ 243

Name_____

 **1·10** ▸ **Compare and Order Money**

Compare. Write >, <, or =.

1. $2.45 _____ $5.42

2. $6.71 _____ $6.17

3. $8.97 _____ $8.97

4. $1.21 _____ $1.12

Write in order from greatest to least.

5. $5.81, $3.20, $4.47

6. $6.31, $6.33, $6.28

_____

_____

7. $2.94, $3.08, $3.11

8. $8.58, $8.88, $8.85

_____

_____

Problem Solving

9. Jorge offers Don $7.62 for a hat he is selling. Katy offers Don the collection of money shown at right. Who offers more?

_____

10. Jodie has 5 dollar bills, 5 nickels, and 5 pennies. Paul has a five-dollar bill, a quarter, and 2 dimes. Who has more money?

_____

Spiral Review

Round to the nearest ten.

11. 88 _____

12. 45 _____

13. 162 _____

14. 286 _____

Name_____

 **2·1** **Addition Properties**

Add. Use mental math.

1. 5 + 3 = _____

2. 17 + 2 = _____

3.   4
  + 5

4.   3
  + 8

5.   7
  + 0

6.   5
  + 9

Find each missing number. Name the addition property you used.

7. 7 + 2 = 2 + d

8. 9 + z = 4 + 9

9. 6 + h = 6

_____

_____

_____

Add. Then use the Commutative Property to write a different addition sentence.

10. 6 + 3

11. 0 + 13

12. 9 + 8

_____

_____

_____

Problem Solving

13.   There are 8 cars sitting in a parking lot. In the next hour,
    9 more cars pull up. How many cars are there total?   _____

14.   In the year 1790, there were 13 states in the United
    States. In the next two years, 2 more states joined.
    How many states were there total?   _____

Spiral Review

Write the value of each underlined digit.

15. 5<u>7</u> _____

16. 1,<u>6</u>14 _____

17. <u>3</u>9,179 _____

 **2·2** **Add 3 or More Numbers**

Add. Show how you used the Associative Property.

1. 2 + 4 + 5 _____

2. 8 + 1 + 6 _____

Use the Commutative and Associative properties to add mentally. Explain what you did.

3. 2 + 3 + 3   **4.** 6 + 1 + 0 + 4   **5.** 4 + 4 + 1 + 5

_____   _____   _____

_____   _____   _____

Find each missing number.

6. (6 + 2) + 4 = *b* + (2 + 4) _____

7. 5 + (*v* + 9) = (5 + 3) + 9 _____

Problem Solving

8. In a school mural, students have painted 7 dogs, 5 cats, and 4 snakes. How many animals are in the mural in all?   _____

9. A table at a rummage sale has 6 jigsaw puzzles, 2 lamps, 3 wool hats, and 9 books for sale. How many items are on the table in all?   _____

Spiral Review

Compare. Write >, <, or =.

10. 78 _____ 97   **11.** 115 _____ 151   **12.** 393 _____ 391

2·3 Addition Patterns

Complete the equations.

1. 2 + 5 = _____
   20 + 50 = _____
   200 + 500 = _____

2. 9 + 9 = _____
   90 + 90 = _____
   900 + 900 = _____

Add. Use mental math.

3. 400 + 80 + 50 = _____

4. 4,000 + 4,000 = _____

5. 2,000 + 5,000 = _____

Problem Solving

6. There are 200 students sitting in the school auditorium. Another 500 students come in. How many students are in the auditorium in all?

   _____

7. Delaware's area is about 2,400 square miles, and Rhode Island's area is about 1,200 square miles. What is their combined area?

   _____

   Source: *The World Almanac and Book of Facts 2000*

Spiral Review

Order from least to greatest.

8. 712; 1,488; 360 _____

9. 4,877; 4,725; 5,220 _____

10. 6,767; 6,679; 6,776 _____

11. 9,939; 9,993; 9,399 _____

2-4 ▶ Explore Regrouping in Addition

Find each sum. You may use models.

1.	53 + 29	**2.**	83 + 41	**3.**	176 + 57	**4.**	267 + 354

5.	812 + 146	**6.**	267 + 267	**7.**	932 + 309	**8.**	848 + 252

9. 15 + 28 = _____   **10.** 83 + 51 = _____   **11.** 270 + 187 = _____

12. 482 + 364 = _____   **13.** 621 + 289 = _____   **14.** 356 + 356 = _____

15. 162 + 69 = _____   **16.** 24 + 899 = _____   **17.** 765 + 81 = _____

Solve.

18. A bucket contains 55 tennis balls. Someone adds 76 more tennis balls to the bucket. How many tennis balls are in the bucket in all?

_____

19. A family drives 237 miles on the first day of their vacation. After stopping for the night, they then drive 477 miles on the second day. How many miles have they driven in all?

_____

Spiral Review

Round each number to the nearest thousand or thousand dollars.

20. $2,256 _____   **21.** 54,590 _____   **22.** 81,473 _____

23. $26,902 _____   **24.** $6,954 _____   **25.** 12,409 _____

Name_____

2·5 Add Whole Numbers

Find each sum.

1. $634
    + $218

2. 767
    + 956

3. $675
    + $675

4. 663
    + 441

5. $3.67
    + $4.85

6. 860
    + 576

7. $8.47
    + $5.93

8. 767
    + 777

9. 457 + 264 = _____

10. $7.94 + $2.54 = _____

11. 680 + 897 = _____

12. 535 + 535 = _____

13. 847 + 619 = _____

14. 777 + 999 = _____

15. $968 + $469 = _____

16. 466 + 675 = _____

17. 926 + 678 = _____

Problem Solving

18. New York has 249 miles of track in its subway system. Moscow has 153 miles of track in its subway system. How many miles of track do they have combined? _____

19. Tim's reading book has 732 pages. His math book has 577 pages. How many pages are there in all? _____

Spiral Review

Write the digit in each place named.

20. 9,272 (tens) _____

21. 7,958 (ones) _____

22. 21,583 (hundreds) _____

23. 73,004 (thousands) _____

24. 845,490 (ten thousands) _____

25. 638,112 (hundred thousands) _____

Name_____

 **2·6** **Estimate Sums**

Estimate each sum.

1. 3,882
    + 4,107

2. 2,274
    + 4,730

3. 3,191
    + 1,815

4. 7,210
    + 498

5. 8,689
    + 797

6.   483
      567
    + 924

7. 2,122
    4,867
    + 881

8. 3,857
    4,150
    + 1,096

9. 21 + 67 = _____

10. 36 + 24 = _____

11. 679 + 124 = _____

12. 411 + 296 = _____

13. 562 + 31 = _____

14. 387 + 72 = ____

15. 693 + 515 = _____

16. 785 + 524 = _____

17. 989 + 477 = _____

Problem Solving

18. For the first showing of a new movie, 923 people
     buy tickets. For the second showing, 665 people
     buy tickets. About how many people in all buy tickets?

     _____

19. Angel Falls in Venezuela is 3,212 feet high.
     Yosemite Falls in California is 2,425 feet high.
     About how many feet high would they be combined?

     _____

Spiral Review

20. 18 − 4 = _____

21. 36 − 9 = _____

22. 45 − 29 = _____

Name _____

Name ___

Estimate or Exact Answer

Solve. Explain why you gave an estimated or exact answer.

1. In a basketball game, the Panthers score 88 points and the Mustangs score 85 points. Together, do they score more than 150 points?

_____

2. Last year, Rick's Used Cars sold 212 pickup trucks, 81 vans, and 147 station wagons. How many vehicles did Rick's sell in all?

_____

Use data from the table for problems 3–4.

3. About how many students responded to the poll?

_____

4. In all, how many students answered either "hot dog" or "nachos"?

_____

Favorite Food to Eat While Watching a Movie

Type of Food	Responses
Hot Dog	117
Popcorn	354
Nachos	192
Other	82
None	168

Spiral Review

Write the value of 9 in each number.

5. 977   6. 23,019   7. 962,445   8. 145,890

_____   _____   _____   _____

2·8 Add Greater Numbers

Add. Check if each answer is reasonable.

1. 4,157
    + 2,612

2. 3,351
    + 5,496

3. 4,927
    + 8,391

4. 7,294
    + 968

5. $84.24
    + 26.37

6. 9,377
    + 927

7. 5,243 + 3,614 = _____

8. $11.63 + $35.26 = _____

9. 4,726 + 857 = _____

10. 2,782 + 3,565 = _____

Problem Solving

11. Over the course of a season, a professional
    basketball player scores 1,867 points. His
    teammate scores 1,364 points. How many
    points do they score in all?   _____

12. A moving truck carries 3,792 pounds of
    furniture in its first load of the day, and
    2,810 pounds in its second load. How many
    pounds does it carry in all?   _____

Spiral Review

Order from greatest to least.

13. $5.97, $4.78, $7.60   14. $2.45, $2.55, $2.41   15. $8.68, $8.86, $8.84

_____   _____   _____

Name_____

Problem Solving: Strategy
Write a Number Sentence

Write a number sentence to solve.

1. There are 17 grapefruit and 21 oranges in the refrigerator. How many pieces of fruit are there in all? _____

   _____

2. There are 15 grown dogs and 37 puppies. What is the total number of dogs? _____

   _____

3. In May, 46 boys and 48 girls were born. How many children in all were born in May? _____

Mixed Strategy Review

Use data from the table for problems 4–5. Write a number sentence to solve.

Planet	Moons
Saturn	18
Mars	2
Neptune	8
Jupiter	16
Earth	1

4. The table lists five planets from our solar system and their moons. Which two planets have a combined total of 34 moons?

   _____

5. Uranus has 16 moons more than Mars does. How many moons does Uranus have? _____

Spiral Review

Find the amount of change. List the coins and bills you might receive.

6. Cost: $0.64. You give: $1._____

7. Cost: $2.41. You give: $5. _____

   _____

Name_____

 **2·10** **Add More Than Two Numbers**

Add. Check for reasonableness.

1.	$3.24	2.	781	3.	$36.32
	4.53		976		4.93
	+ 1.22		+ 291		+ 5.65

4.	2,346	5.	937	6.	6,822
	3,520		484		760
	+ 1,976		833		43
			+ 926		+ 1,932

7. $4.18 + $5.91 + $9.33 = _____   8. 3,257 + 1,543 + 2,287 = _____

9. $2.90 + $5.39 + $1.92 = _____   10. 312 + 766 + 881 + 562 = _____

Problem Solving

11. At the drug store, Dawn pays $4.29 for a package
of batteries, $2.95 for a magazine, and $0.67 for a
granola bar. How much does she pay in all? _____

12. The three smallest states in the United States
are Rhode Island, Delaware, and Connecticut.
Rhode Island has an area of 1,231 square miles.
Delaware has an area of 2,396 square miles.
Connecticut has an area of 5,544 square miles.
What is the area of all three states combined? _____
Source: *The World Almanac and Book of Facts 2000*

Spiral Review

Round to the nearest ten or ten dollars.

13. 72 _____   14. $819 _____   15. 945 _____

Name_____

3·1 Relate Addition and Subtraction

Write a group of related sentences for each group of numbers.

1. 3, 13, 16

_____

_____

_____

_____

2. 18, 7, 25

_____

_____

_____

_____

Find each sum or difference. Write a related addition or subtraction fact.

3. 9 + 5 _____

4. 17 − 3 _____

Write the missing numbers in each fact family.

5. 5 + 2 = r _____

5 + r = 7 _____

r − 2 = 5 _____

7 − r = 2 _____

6. 9 + x = 16 _____

x + 7 = 16 _____

16 − x = 9 _____

16 − x = 7 _____

Problem Solving

7. At a bake sale, Hector sets up a table with 12 pies. At the end of the sale, he has 3 pies left. How many pies did he sell?

Spiral Review

Compare. Write >, <, or =.

8. $4.17 _____ $4.71   **9.** $32.16 _____ $31.20   **10.** $87.24 _____ $87.42

Name_____

Identify Extra Information

Solve. Identify the extra information.

1. There are 11 cats staying at a kennel. There are 13 black dogs and 6 yellow dogs at the kennel. How many dogs are there in all?

    _____

2. At a concert, a singer performs for 24 minutes. A total of 352 people are there. The first six songs take up 20 minutes. How much time is left in the performance?

    _____

3. Victor is supposed to swim 18 laps in the pool, but he gets tired and stops with 3 laps to go. The pool is 50 meters long. How many laps does he swim?

    _____

Use data from the table for problems 4–5.

4. There are 5 more pennies than dimes in Leah's coin jar. How many more nickels than dimes are in the jar?

    _____

5. How many silver-colored coins are there in all?

    _____

Leah's Coin Jar

Coin	Number
Quarter	16
Dime	9
Nickel	21
Penny	14

Spiral Review

Round to the nearest ten or ten dollars.

6. 39 _____   7. $273 _____   8. 445 _____   9. $164 _____

Name_____

 **3·3** ## Subtraction Patterns

Write the number that makes each sentence true.

1. $9 - 5 = d$ _____

$90 - 50 = e$ _____

$900 - 500 = f$ _____

2. $7 - g = 5$ _____

$70 - h = i$ _____

$700 - i = 500$ _____

Subtract mentally.

3. 1,200
     $-$ 300

4. 505
     $-$ 300

5. 170
     $-$ 80

6. 754
     $-$ 200

7. $60 - 30 =$ _____

8. $1,600 - 800 =$ _____

Problem Solving

9. The Mississippi runs through or along the borders of 10 of the 50 United States. How many states don't touch the river at all? _____

10. There are 2,000 books overdue at the local library. After the library sends out notices, 1,200 of the books are brought back. How many books are still overdue? _____

Spiral Review

Write the digit in the ten thousands place.

11. 264,055 _____

12. 772,636 _____

13. 100,320 _____

14. 647,647 _____

3·4 Explore Regrouping in Subtraction

Find each difference. You may use models.

1. 45
  – 8
	2. 64
	  – 37
		3. 77
		  – 29

4. 136
  – 29
	5. 351
	  – 122
		6. 946
		  – 527

7. 429
  – 162
	8. 740
	  – 327
		9. 532
		  – 148

10. 93 – 4 = _____

11. 48 – 29 = _____

12. 181 – 78 = _____

13. 827 – 209 = _____

14. 680 – 356 = _____

15. 925 – 471 = _____

Solve.

16. In a basketball game, Erin scores 18 points. Her team, the Gila Monsters, scores 67 points in all. How many points does the rest of the team score? _____

17. Hutchison Hill is 841 feet high. There is a picnic shelter on the hill, located 382 feet below the top. How high up is the picnic shelter? _____

Spiral Review

Add. Then use the Commutative Property to write a different addition sentence.

18. 6 + 2

_____

19. 3 + 8

_____

Name_____

 **3·5** ## Subtract Whole Numbers

Subtract. Remember to check your answer.

1. 472
    − 125

2. 893
    − 268

3. 721
    − 493

4. $6.21
    − 1.63

5. 485
    − 106

6. 371 − 117 = _____

7. $625 − $458 = _____

8. 952 − 381 = _____

9. $5.45 − $1.89 = _____

Pick pairs of numbers from the set below that have a difference closest to the given target number.

834   442   882   556   134   108   285

10. Target number: 700

11. Target number: 600

12. Target number: 100

_____

_____

_____

Problem Solving

13. A cheetah can run 105 kilometers an hour, and a horse can run 69 kilometers an hour. How much faster can a cheetah run?

_____

14. A movie theater holds 766 people, and 298 have showed up so far. How many seats are left?

_____

Spiral Review

Add. Use mental math.

15. 600 + 200 = _____

16. 700 + 900 = _____

© McGraw-Hill School Division

Name_____

Subtract. Add to check.

1. 208
   − 47

2. 607
   − 39

3. 405
   − 217

4. $605
   − 315

5. 700
   − 447

6. $900
   − 261

7. 803
   − 294

8. 901
   − 109

9. 102 − 28 = _____

10. $709 − $263 = _____

11. 504 − 160 = _____

12. 300 − 79 = _____

13. $800 − $426 = _____

14. 607 − 209 = _____

Problem Solving

15. Ramon's class is trying to collect 100 glass bottles for a project on recycling. They need 47 more bottles to meet their goal. How many bottles have they collected so far? _____

16. Paige and four of her friends step on a scale together as a joke. The scale reads 309 pounds. Paige steps off, and the scale reads 237 pounds. How many pounds does Paige weigh? _____

Spiral Review

Find each missing number. Name the addition property you used.

17. $7 + 6 = 6 + x$

18. $9 + t = 9$

_____

_____

Name_____

3·7 Estimate Differences

Name

Estimate each difference.

1. 78 – 21 _____

2. 817 – 380 _____

3. 1,488 – 279 _____

4. 429 – 51 _____

5. 2,492 – 612 _____

6. 4,326 – 895 _____

7. 5,976 – 4,180 _____

8. 7,098 – 2,231 _____

9. 6,361 – 1,470 _____

10. 4,234 – 871_____

11. 8,992 – 2,009_____

12. 6,647 – 3,166_____

Problem Solving

13. There are 79 people waiting in line to buy
tickets for a concert. When the tickets are
all gone, there are still 31 people waiting
in line. About how many people were able
to buy tickets?   _____

14. There are 3,929 beans in a sack. Someone
spills 1,124 of the beans. About how many
beans are left in the sack?   _____

Spiral Review

Complete.

15. $4 + 3 = f$ _____

    $40 + 30 = g$ _____

    $400 + 300 = h$ _____

16. $7 + c = 13$ _____

    $70 + d = 130$ _____

    $700 + e = 1,300$ _____

3·8 Problem Solving: Strategy
Write a Number Sentence

Write a number sentence to solve.

1. Don has enough buns for 120 hamburgers. He makes 73 hamburgers and serves them on buns. How many buns does he have left?

   _____

2. Taft Elementary School has 383 students. McKinley Elementary School has 635 students. How many more students does McKinley have?

   _____

3. The NCNB Building in Charlotte, is 871 feet tall. The John Hancock Center in Chicago is 256 feet taller. How tall is the John Hancock Center?

   _____

Use data from the table for problems 4–5. Write a number sentence to solve.

4. How many sausage pizzas and vegetable pizzas were sold in all?

   _____

October Pizza Sales

Pizza	Number Sold
Pepperoni	573
Sausage	219
Cheese	295
Vegetable	377

5. How many more pepperoni pizzas than cheese pizzas were sold?

   _____

Spiral Review

Add. Show how you used the Associative Property.

6. 4 + 1 + 3 _____

7. 9 + 2 + 4 _____

Name_____

3·9 Subtract Greater Numbers

Subtract. Check that each answer is reasonable.

1. 5,286
   − 2,171

2. 3,649
   − 1,583

3. 7,922
   − 3,482

4. 9,373
   − 2,655

5. $52.27
   − 13.82

6. 1,754
   − 476

7. $83.46
   − 9.22

8. 6,003
   − 1,298

9. $46.09
   − 7.22

10. $42.55 − $2.72 = _____

11. 7,964 − 6,543 = _____

12. 9,267 − 9,265 = _____

13. $78.23 − $15.66 = _____

14. 4,001 − 3,902 = _____

15. 8,070 − 3,735 = _____

Problem Solving

16. At a sale, $7.39 is subtracted from the price of a
    chair that usually costs $73.90. What is the sale
    price of the chair?   _____

17. An adult hippopotamus weighs 5,512 pounds,
    and an adult giraffe weighs 3,527 pounds.
    How much more does the hippopotamus weigh?   _____

Spiral Review

Find each sum.

18. 738
   + 97

19. 560
   + 384

20. 459
   + 732

21. $8.18
   + $6.72

Name_____

 **4·1** **Tell Time**

Write each time using A.M. or P.M. Then write one way to read each time.

1.

_____

_____

2.

_____

_____

Problem Solving

Write each time using A.M. or P.M. Tell how you know.

3. _____

_____

4.
_____

_____

Spiral Review

Estimate each sum.

5. 689 + 216 _____   6. 4,291 + 493 _____

Name_____

4-2 Convert Time

Tell how much time.

1. 3 quarter hours = _____ minutes

2. 1 hour = _____ half hours

3. 60 minutes = _____ quarter hours

4. $\frac{1}{4}$ hour = _____ minutes

5. 30 minutes = _____ hour

6. 2 hours = _____ quarter hours

Problem Solving

7. Johanna waits for the bus for 30 minutes. How long does she wait in quarter hours? _____

8. The night before a test, Miguel spends 120 minutes studying. How many hours does he study? _____

9. The space flight of Yuri Gagarin, the first man in space, was a little over 7 quarter hours long. About how many minutes did his flight last?

_____

Spiral Review

Find each sum or difference. Write a related addition or subtraction fact.

10. 7 − 3 = _____   11. 6 + 5 = _____   12. 16 − 7 = _____

_____   _____   _____

Name_____

How much time has passed?

1. Begin: 7:20 A.M. End: 7:40 A.M.

_____

2. Begin: 3:15 P.M. End: 9:50 P.M.

_____

3. Begin: 9:40 P.M. End: 2:30 A.M.

_____

4. Begin 11:23 A.M. End: 8:41 P.M.

_____

What time will it be in 3 hours?

5.  _____

6. _____

Problem Solving

7. Julian leaves his house at 6:45 A.M. for a 35-minute run. What time does he finish running? _____

8. Teresa goes to a barbecue and gets back home exactly 4 hours after she left. She gets home at 5:08 P.M. What time did she leave home? _____

9. It takes 100 minutes for Elaine to drive from her house to visit her grandparents. She starts driving at 4:30 P.M. Will she be there by 6:00 P.M.? _____

Spiral Review

Add. Check if each answer is reasonable.

10. 5,205
    + 2,721
    ———

11. 1,829
    + 3,166
    ———

12. $37.45
    + $44.87
    ———

© McGraw-Hill School Division

Name_____

4·4 Calendar

Use the calendar to solve problems 1–3.

1. Antonio plays in a band that only performs on the fourth Monday of every month. Will he perform on Memorial Day?

   _____

2. The band rehearses every Wednesday. How many times will it rehearse in May?

   _____

3. How many weeks is it from May 10 to Memorial Day?   _____

M						
Sunday	Monday	Tuesday	Wed.			
						1
2	3	4	5	6	7	8
9	10	11	12	13	14	15
16	17	18	19	20	21	22
23	24	25	26	27	28	29
30	31 Memorial Day					

Problem Solving

4. City elections are held on November 4. The winners begin their new jobs three weeks later. On what date do they begin work?

   _____

5. The mayor's term lasts four years. How many months is that?

   _____

Spiral Review

Find the missing number.

6. $9 - x = 2$ _____   7. $15 - r = 7$ _____   8. $m - 6 = 13$ _____

Identify Missing Information

1. Luis gets out of school at 3:30 P.M., goes straight home, and walks his family's dog for an hour. He finishes walking the dog at 5:00. How long does it take him to get home? _____

2. Kirsten started getting ready for bed at 8:30 P.M. She slept for 10 hours and woke up the next morning at 7:15 A.M. How long did it take her to get ready for bed? _____

3. Terrell's family goes to see an early showing of a movie, which starts at 1:25 P.M. It takes his family 20 minutes to leave the theater and drive home. They return home at 3:50 P.M. How long is the movie? _____

Use data from the bus schedule for problems 4–5.

Destination	Depart	Arrive
Springfield	12:15 P.M.	2:30 P.M.
Fairview	1:35 P.M.	3:15 P.M.
Moose City	4:07 P.M.	8:58 P.M.
Dos Rios	5:22 P.M.	7:11 P.M.

4. How long is the bus trip to Springfield if you depart at 12:15?

_____

5. How long after the bus to Moose City leaves does the bus to Dos Rios

arrive at its destination? _____

Spiral Review

Subtract. Use mental math.

6. 170 − 130 = _____   7. 640 − 60 = _____   8. 853 − 143 = _____

Name_____

4·6   Problem Solving: Strategy
Work Backward

Work backward to solve.

Name_____

1. On Friday morning, Carlos goes out for a run that lasts 30 minutes. When he gets back home, he sees the clock at the right. What time did he start running?

   _____

2. The school cafeteria has 74 apples left after two classes have eaten. At the start of the day the cafeteria had 120 apples, and the first class to come through took 22 apples. How many apples did the second class take?

   _____

Mixed Strategy Review

Use data from the table for problem 3.

February Snowfall in Cedarville	
Storm #1	366 mm
Storm #2	218 mm
Storm #3	—
Total	871 mm

3. The table above lists the amount of snow, in millimeters, that fell in Cedarville during the month of February. Unfortunately, someone forgot to fill in the amount of snow that fell during the third snowstorm. You know that 871 millimeters of snow in all fell during the month. How much snow fell during the third storm?

   _____

Spiral Review

4. 747
    − 236

5. 839
    − 56

6. 465
    − 157

© McGraw-Hill School Division

4·7 Line Plots

Make a tally chart and a line plot.

1. Use the data below to make a tally chart and line plot. The number of pets owned by each student in Ms. Green's class is as follows: 0, 3, 1, 1, 2, 0, 3, 0, 2, 4, 1, 1, 4, 2, 2, 0, 1, 2, 0, 3, 1, 1, 2, 2, 0, 2

Use the information from the chart for problems 2–3.

2. How many students have 2 pets? _____

3. How many more students have 1 pet than have 4 pets? _____

Problem Solving

Use data from the line plot for problems 4–5.

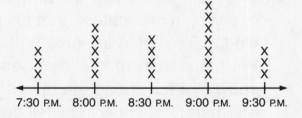

4. At which two times do the fewest students go to bed?

_____

5. From 8:00 P.M. to 9:00 P.M., how many students go to bed?

_____

Spiral Review

6. 806
   − 282

7. $7.07
   − $5.48

8. 500
   − 147

4·8 Explore Pictographs

Use data from the pictograph for problems 1–4.

1. How many visitors came to the center on Wednesday?

_____

2. On which day did the fewest visitors come to the center?

_____

3. How many visitors came over the weekend?

_____

Visitors to the Wildlife Center
June 23–June 29

Day	Number of Visitors
Monday	🧍🧍🧍🧍🧍
Tuesday	🧍🧍🧍
Wednesday	🧍🧍🧍🧍
Thursday	🧍🧍🧍
Friday	🧍🧍🧍
Saturday	🧍🧍🧍🧍🧍🧍🧍
Sunday	🧍🧍🧍🧍🧍

Key: Each 🧍 stands for 10 visitors

4. How many more visitors came to the center on Monday than came on Friday? _____

Solve.

5. The next Friday, 5 visitors stop by before lunch, and 35 come in the afternoon. How would you write the total number of visitors in the pictograph?

_____

6. The next Sunday is sunny and there are two times as many visitors as on the Sunday before. How would you write this total in the pictograph?

_____

Spiral Review.

7. 319
    + 467

8. 644
    − 219

9. 852
    + 587

10. 852
     − 587

Name_____

 **4·9** **Explore Bar Graphs**

Use data from the bar graph for problems 1–3.

1. How many students have their parents drive them to school?

   _____

2. What is the most common way for the students in the class to get to school? How many students get to school this way?

   _____

3. How many students either bicycle or walk to school?

   _____

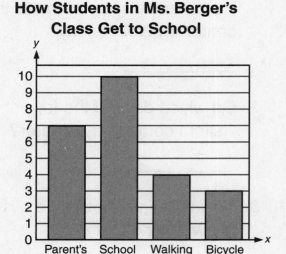

How Students in Ms. Berger's Class Get to School

Solve. Fill in the bar graph at right for problem 4.

4. In Ms. Berger's class, 9 students have to take out the garbage, 6 have to clean the bathroom, and 7 have to unload the dishwasher. Fill in the bar graph to show these numbers.

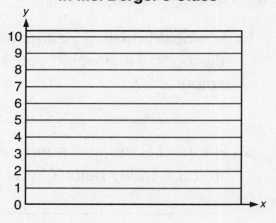

Household Chores for Students in Ms. Berger's Class

Spiral Review

Write the value of each underlined digit.

5. 4<u>5</u>,241

6. 733,<u>7</u>90

7. 202,4<u>1</u>6

8. <u>3</u>46,920

_____   _____   _____   _____

Name_____

4·10 Coordinate Graphing

Use this map of the Elm Creek Mall to answer problems 1–10.

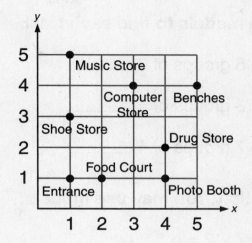

Use the map to write the location.

1. Entrance _____

2. Drug Store _____

3. Benches _____

4. Computer Store _____

Name the place at each location on the map.

5. (1,5) _____   6. (2,1) _____

7. (4,1) _____   8. (1,3) _____

Problem Solving

9. If it takes you 3 minutes to walk from the Music Store to the Shoe Store, how long will it take to walk from the Shoe Store to the Entrance? _____

10. If you start at (3,4) and walk to (4,2), what store do you begin at? What store do you walk to? _____

Spiral Review

11.	7,394	12.	5,467	13.	6,252	14.	8,421
	− 3,214		− 3,592		− 784		− 4,348

Name_____

5·1 ▶ Explore the Meaning of Multiplication

Use models to find each total.

1. 5 groups of 6 _____   **2.** 9 groups of 2 _____

3. 3 groups of 5 _____   **4.** 6 groups of 6 _____

5. 7 groups of 4 _____   **6.** 2 groups of 8 _____

Multiply. You may use models.

7. $7 \times 2 =$ _____   **8.** $2 \times 4 =$ _____

9. $5 \times 3 =$ _____   **10.** $4 \times 6 =$ _____

11. $2 \times 8 =$ _____   **12.** $5 \times 4 =$ _____

13. $3 \times 7 =$ _____   **14.** $4 \times 4 =$ _____

Solve.

15. While exploring a creek, 6 children each pick
up 3 rocks to take home with them. How many
rocks do they take in all?   _____

16. In each of six successful moon missions, two
astronauts walked on the moon's surface. How
many astronauts have walked on the moon?   _____

Spiral Review

Round each number to the nearest hundred.

17. 478 _____   **18.** 7,561 _____   **19.** 1,782 _____   **20.** 3,245 _____

Name_____

 **5-2** ▶ **Relate Multiplication and Addition**

Find each total. Write an addition sentence and a multiplication sentence.

1.

2.

_____   _____

_____   _____

Multiply.

3. $2 \times 4 =$ _____   4. $3 \times 3 =$ _____   5. $6 \times 2 =$ _____

6. $5 \times 5 =$ _____   7. $7 \times 3 =$ _____   8. $4 \times 6 =$ _____

Describe and complete each skip-counting pattern.

9. 6, 12, 18, 24, 30, _____   10. 9, 18, _____, 36, 45

_____   _____

Problem Solving

11. A bookcase has 3 shelves, each of which has 9 books on it. How many

books are in the bookcase in all? _____

12. Graciela finds 5 nickels. How much money does she find? _____

Spiral Review

13. $445 + 352 =$ _____   14. $831 + 584 =$ _____

Name_____

Write the multiplication sentence each array shows.

1. ⚫⚫⚫⚫⚫⚫⚫
   ⚫⚫⚫⚫⚫⚫⚫

   _____

2. ⚫⚫⚫⚫⚫
   ⚫⚫⚫⚫⚫
   ⚫⚫⚫⚫⚫

   _____

Find each product. Then use the Commutative Property to write a different multiplication sentence.

3. $3 \times 7 =$ _____

4. $4 \times 2 =$ _____

5. $8 \times 1 =$ _____

6. $9 \times 3 =$ _____

7. $5 \times 6 =$ _____

8. $2 \times 8 =$ _____

Solve.

9. A standard egg carton can be looked at as an array of 2 rows of 6 eggs each. How many eggs are in a standard egg carton? _____

10. Half of a checkerboard can be looked at as an array of 4 rows of 8 squares each. How many squares are in one half of a checkerboard?

_____

Spiral Review

How much time has passed?

11. Begin: 12:15 A.M. End: 12:45 A.M.   12. Begin: 3:15 P.M. End: 4:35 P.M.

_____   _____

Name_____

5·4 Problem Solving: Reading for Math
Choose an Operation

Solve. Tell how you chose the operation.

1. At a garage sale, Ellis buys 7 cassettes and Tom buys 3 cassettes. How many cassettes do they buy in all? _____

_____

2. Each of the three cassettes that Tom buys has 9 songs on it. How many songs are on the cassettes in all? _____

_____

3. At the same sale, Ellis, Tom, Denise, and Kyra each buy 4 postcards. How many postcards do they buy all together? _____

_____

Use data from the table for problems 4–6.

4. If you order 3 medium pizzas, how many slices do you get in all?

_____

5. How many more slices are in a large pizza than are in a small pizza?

_____

Pizza Size	Number of Slices
Small	6
Medium	8
Large	12

6. If you order one pizza of each size, how many slices do you get in all?

_____

Spiral Review

7. 287 − 129 = _____   8. 862 − 468 = _____   9. 905 − 275 = _____

5·5 Multiply by 2 and 5

Find each product.

1. $\begin{array}{r} 6 \\ \times\,2 \\ \hline \end{array}$	**2.** $\begin{array}{r} 3 \\ \times\,5 \\ \hline \end{array}$	**3.** $\begin{array}{r} 7 \\ \times\,2 \\ \hline \end{array}$
4. $\begin{array}{r} 8 \\ \times\,5 \\ \hline \end{array}$	**5.** $\begin{array}{r} 9 \\ \times\,2 \\ \hline \end{array}$	**6.** $\begin{array}{r} 4 \\ \times\,5 \\ \hline \end{array}$

7. $2 \times 4 =$ _____ **8.** $5 \times 7 =$ _____ **9.** $6 \times 5 =$ _____

10. $5 \times 2 =$ _____ **11.** $2 \times 8 =$ _____ **12.** $9 \times 5 =$ _____

Find each total.

13. $(2 + 1) \times 2 =$ _____ **14.** $(2 + 3) \times 5 =$ _____

15. $(1 + 1) \times 7 =$ _____ **16.** $(3 + 2) \times 4 =$ _____

Problem Solving

17. Ines notices that each of the 3 trees in front of her house has exactly 5 branches. How many branches do the trees have all together?

_____

18. There are 8 bicycles leaning up against a fence. How many wheels are there in all? _____

Spiral Review

Compare. Write >, <, or =.

19. $4.92 ____ $40.91 **20.** $16.32 ____ $16.27 **21.** $80.29 ____ $82.09

5·6 Multiply by 3 and 4

Multiply.

1. 2
   × 3

2. 4
   × 3

3. 3
   × 6

4. 2
   × 4

5. 3
   × 7

6. 4
   × 8

7. 2 × 5 = _____

8. 4 × 4 = _____

9. 4 × 1 = _____

10. 3 × 3 = _____

11. 6 × 4 = _____

12. 5 × 3 = _____

Complete the skip counting pattern.

13. 18, 21, 24, 27, 30 _____

14. 16, 20, 24, _____, 32, 36

Problem Solving

15. In the window of a music store, there are 4 guitars for sale. Each guitar

   has 6 strings. How many strings are there in all? _____

16. An apartment building 3 floors high has 7 apartments on each floor.

   How many apartments are in the building in all? _____

Spiral Review

Order from least to greatest.

17. 738, 680, 751

18. 320, 322, 302

19. 932, 923, 929

_____   _____   _____

5·7 **Problem Solving: Strategy**

Find a Pattern

Find a pattern to solve.

1. There are 16 cars parked in a line in a parking lot. Every fourth car is dark blue. All the others are red. What color is the 12th car in the line?

   _____

2. A grocery store clerk puts 18 cans in the first row of a display. In the second row she puts 16 cans, and in the third row she puts 14 cans. If this pattern continues, how many cans will the sixth row have?

   _____

3. Flyaway Airlines is offering a sale on tickets from St. Louis to Dallas. On the first day of the sale, a ticket costs $330. On the second day the price is $300, and on the third day the price is $270. If this pattern continues, how much will a ticket cost on the seventh day of the sale?

   _____

Mixed Strategy Review

4. The stripes on an American flag begin with a red stripe, and then alternate red and white for 13 stripes. What color is the 8th stripe? _____

5. Trains have left the Kernersville station at 12:07 P.M., 2:07 P.M., and 4:07 P.M. If this pattern continues, what time will the next train leave? _____

Spiral Review

6.	802	7.	462	8.	273	9.	2,005
	− 457		+ 439		+ 849		− 1,362

5·8 Multiply 0 and 1

Find each product.

1. 3
    ×1

2. 7
    ×0

3. 1
    ×5

4. 0
    ×6

5. 8
    ×1

6. 2
    ×0

7. $1 \times 3 =$ _____

8. $9 \times 0 =$ _____

9. $6 \times 1 =$ _____

Find each missing number.

10. $3 \times$ _____ $= 3$

11. _____ $\times 7 = 0$

12. $1 \times$ _____ $= 8$

Problem Solving

13. Jane is driving down the highway at 63 miles per hour. At this speed, how far will she travel in one hour? How can you tell?

_____

14. Doug can run 243 meters in one minute. On Sunday, he reads all day and runs for 0 minutes. How far does Doug run on Sunday?

_____

Spiral Review

Write each number in expanded form.

15. 497

16. 3,446

_____   _____

5·9 Multiplication Table

Use the multiplication table to solve.

0	0	0	0	0	0	0	0	0	0	0
1	0	1	2	3	4	5	6	7	8	9
2	0	2	4	6	8	10	12	14	16	18
3	0	3	6	9	12	15	18	21	24	27
4	0	4	8	12	16	20	24	28	32	36
5	0	5	10	15	20	25	30	35	40	45
6	0	6	12	18	24	30	36	42	48	54
7	0	7	14	21	28	35	42	49	56	63
8	0	8	16	24	32	40	48	56	64	72
9	0	9	18	27	36	45	54	63	72	81
	0	1	2	3	4	5	6	7	8	9

1. Which rows have mixed odd and even numbers? _____

2. Which columns have mixed odd and even numbers? _____

Compare. Write >, <, or =.

3. 4×3 ____ 4×2 4. 7×3 ____ 3×8 5. 5×0 ____ 2×2

Problem Solving

6. To get to a game, Jeni's soccer team goes in 4 cars, each carrying 4 players. How many players are on her team? _____

7. There are 7 tables in a restaurant, and 3 customers are sitting at each table. How many customers are in the restaurant? _____

Spiral Review

Write the money amount.

8.

9.

_____   _____

 **6·1** ▶ **Explore Square Numbers**

Write a multiplication sentence.

1.  _____

2. 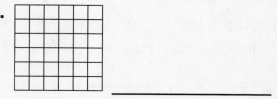 _____

Draw the model and find the product.

3. $4 \times 4 =$ _____

4. $2 \times 2 =$ _____

Solve.

5. A square chart has 7 rows of boxes and 7 columns of boxes. How many boxes are in the chart?   _____

6. Anna's muffin box has one layer of muffins in 5 rows and 5 columns. How many muffins are in the box?   _____

Spiral Review

For each group of numbers, write a group of related addition and subtraction sentences.

7. 5, 9, 14

_____

_____

_____

8. 16, 7, 23

_____

_____

_____

Name_____

6-2 Problem Solving: Reading for Math
Solve Multistep Problems

Solve. Tell what hidden questions you inferred.

1. At the bookstore, Lori buys 2 new books for $6 each and 4 used books for $4 each. How much does she spend at the bookstore all together?

   _____

   _____

2. In his garden, Jeff has 3 rows of 5 tomato plants each and 2 rows of 8 carrot plants each. How many plants are in Jeff's garden?

   _____

   _____

Use data from the table for problems 3–4.

3. The table shows the prices of fruit for sale at the store. Ben wants to buy 2 pounds of mangoes and a honeydew melon. How much will it cost him?

   _____

Fruit	Price
Basket of Blueberries	$2
Basket of Raspberries	$3
Honeydew Melon	$5
1 pound of Mangoes	$6

4. Anita buys 5 baskets of blueberries, and Jon buys 4 baskets of raspberries. Who spends more? How much more?

   _____

Spiral Review

5. 89
   − 27

6. 112
   − 75

7. 736
   − 252

8. 866
   − 688

© McGraw-Hill School Division

Name_____

6·3 Multiply by 6 and 8

Write a multiplication sentence for each picture.

1.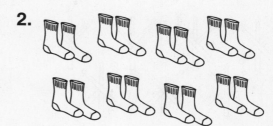

2.

_____   _____

Find each product.

3.   6
   × 4

4.   8
   × 4

5.   7
   × 6

6. 8 × 7 = _____

7. 6 × 5 = _____

8. 1 × 6 = _____

Problem Solving

9. Dennis's dad bought 6 bags of cinnamon-raisin bagels. Each bag has 5 bagels in it. How many bagels did Dennis's dad buy in all?

   _____

10. At the Olympic games, 3 medals are usually given out for each event: one for first place, one for second place, and one for third place. After 8 events, how many medals will have been given out?

   _____

Spiral Review

11.   2
    × 7

12.   5
    × 4

13.   7
    × 5

14.   2
    × 9

Name _____

 **6·4** **Multiply by 7**

Write a multiplication sentence for each picture.

1.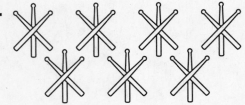

2.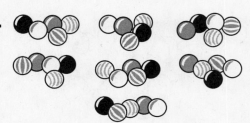

_____   _____

Find each product.

3. $\begin{array}{r} 7 \\ \times\,2 \\ \hline \end{array}$

4. $\begin{array}{r} 7 \\ \times\,6 \\ \hline \end{array}$

5. $\begin{array}{r} 1 \\ \times\,7 \\ \hline \end{array}$

6. $4 \times 7 =$ _____

7. $7 \times 5 =$ _____

8. $7 \times 7 =$ _____

Problem Solving

9. There are 7 boxes of shoes on the bargain table. Each box contains 2 shoes. How many shoes are on the bargain table in all?

10. After Pamela breaks her arm, the doctor tells her she'll have to wear a cast for 6 weeks. How many days will that be?

_____   _____

Spiral Review

What time will it be in 2 hours?

11. _____

12.  _____

6·5 Multiply by 10

Find each product.

1. 10
× 2

2. 10
× 6

3. 10
× 5

4. 7
× 10

5. 10
× 1

6. 9
× 10

7. 10
× 3

8. 4
× 10

9. $10 \times 8 =$ _____

10. $10 \times 0 =$ _____

11. $5 \times 10 =$ _____

12. $10 \times 7 =$ _____

13. $8 \times 5 =$ _____

14. $7 \times 4 =$ _____

Problem Solving

15. Dan has a lawn mowing business and earns $10 for every lawn he mows. If he mows 5 lawns over the weekend, how much money does he make?

_____

16. There are 10 racks in the science lab, each of which holds 6 test tubes. If all of the racks are full, how many test tubes are in the science lab?

_____

Spiral Review

17. 2,453
+ 4,326

18. 5,263
+ 2,927

19. 7,640
+ 891

20. 2,658
+ 1,594

Name_____

 **6·6** ▶ **Multiply by 9**

Write a multiplication sentence for each picture.

1.

_____

2.

_____

Find each product.

3. 9
   × 4

4. 9
   × 2

5. 5
   × 9

6. 3 × 9 = _____

7. 9 × 6 = _____

8. 9 × 9 = _____

Problem Solving

9. At the beginning of a baseball game, the managers for each team announce their starting lineups. Each lineup has 9 players. How many players total are in the starting lineups? _____

10. Kevin's baby sister goes through 9 diapers every day. How many diapers will she go through in a week? _____

Spiral Review

Complete.

11. 4 quarter hours = ____ half hours   12. 2 hours = ____ minutes

Name_____

6-7 Multiplication Table

Use the multiplication table to find each product.

1. $4 \times 3 =$ _____

2. $6 \times 3 =$ _____

3. $8 \times 12 =$ _____

4. $11 \times 10 =$ _____

5. $12 \times 11 =$ _____

0	0	0	0	0	0	0	0	0	0	0	0	0
1	0	1	2	3	4	5	6	7	8	9	10	11	12
2	0	2	4	6	8	10	12	14	16	18	20	22	24
3	0	3	6	9	12	15	18	21	24	27	30	33	36
4	0	4	8	12	16	20	24	28	32	36	40	44	48
5	0	5	10	15	20	25	30	35	40	45	50	55	60
6	0	6	12	18	24	30	36	42	48	54	60	66	72
7	0	7	14	21	28	35	42	49	56	63	70	77	84
8	0	8	16	24	32	40	48	56	64	72	80	88	96
9	0	9	18	27	36	45	54	63	72	81	90	99	108
10	0	10	20	30	40	50	60	70	80	90	10	11	12
11	0	11	22	33	44	55	66	77	88	99	110	121	132
12	0	12	24	36	48	60	72	84	96	108	120	132	144
	0	1	2	3	4	5	6	7	8	9	10	11	12

Problem Solving

6. At a city park, there are 4 soccer fields. On each field, 12 people are playing soccer. How many people are playing soccer in all? _____

7. Eric is trying to cut a piece of paper into 100 pieces. If he cuts it into 11 rows and 9 columns, will he do it? How can you tell?

_____

Spiral Review

Find each product. Then use the Commutative Property to write a different multiplication sentence.

8. $7 \times 3 =$ _____   **9.** $2 \times 9 =$ _____   **10.** $7 \times 6 =$ _____

_____   _____   _____

Name_____

 **6·8** **Problem Solving: Strategy**

Make an Organized List

Use an organized list to solve.

1. Talisa and her family go to a restaurant. Talisa has a choice between a hamburger and a chicken sandwich for her main course, and between French fries and rice as a side dish. Her parents tell her she has to drink milk with her meal. How many different meal combinations does she have? _____ What are they?

_____   _____

_____   _____

2. If Talisa can choose between juice and milk for her drink, how many different meal combinations does she have? _____ What are they?

_____   _____

_____   _____

_____   _____

Mixed Strategy Review

3. Karen is flying from Chicago to San Diego. Her plane has 35 rows with 8 seats in each row. If every seat is taken, how many passengers will there be? _____

Spiral Review

Write the missing numbers in each fact family.

4. $8 + 1 = x$ _____   5. $9 + h = 14$ _____
   $1 + x = 9$ _____ $5 + 9 = h$ _____
   $9 - x = 8$ _____ $h - 5 = 9$ _____
   $x - 8 = 1$ _____ $14 - h = 5$ _____

Name_____

 **6·9** **Multiply Three Numbers**

Find each product.

1. $2 \times 2 \times 3 =$ _____

2. $4 \times 7 \times 2 =$ _____

3. $5 \times 3 \times 3 =$ _____

4. $7 \times 1 \times 2 \times 5 =$ _____

5. $(2 \times 4) \times 9 =$ _____

6. $3 \times 4 \times 5 \times 2 =$ _____

7. $2 \times 2 \times 2 \times 2 =$ _____

8. $3 \times 2 \times 1 \times 8 =$ _____

9. $9 \times 2 \times 0 \times 6 =$ _____

10. $6 \times 2 \times 5 \times 2 =$ _____

Problem Solving

11. On a display table at the grocery store, there are 5 rows and 3 columns of bags. Inside each bag are 4 rolls. How many rolls are on the table all together? _____

12. The flag of Colombia is made up of 3 stripes: one yellow, one red, and one blue. Fred's Flags receives a shipment of 2 boxes, each of which has 7 Colombian flags in it. How many stripes are there on all the flags combined? _____

Spiral Review

Use mental math.

13. $\begin{array}{r} 180 \\ -\ 70 \\ \hline \end{array}$

14. $\begin{array}{r} 1,500 \\ -\ 900 \\ \hline \end{array}$

15. $\begin{array}{r} 1,305 \\ -\ 600 \\ \hline \end{array}$

Name_____

Use counters to complete.

	Total	Number of groups	Number in each group
1.	10	5	
2.	18		6
3.	9	3	
4.	14	2	
5.	12		4
6.	25		5
7.	30	10	
8.	24		6
9.	27	3	
10.	32		4

Solve.

11. There are 15 students in 3 equal groups. How many students are in each group? _____

12. If 24 desks are arranged in 4 equal rows. How many desks are in each row? _____

Spiral Review

13. $4 \times 5 =$ _____   **14.** $7 \times 3 =$ _____   **15.** $9 \times 6 =$ _____

Name_____

7·2 Division as Repeated Subtraction

Write the division sentence.

1. $10 - 5 = 5; 5 - 5 = 0$

_____

2. $9 - 3 = 6; 6 - 3 = 3; 3 - 3 = 0$

_____

Divide.

3. $12 \div 2 =$ _____

4. $8 \div 2 =$ _____

5. $18 \div 3 =$ _____

6. $24 \div 8 =$ _____

7. $20 \div 5 =$ _____

8. $12 \div 4 =$ _____

9. $32 \div 4 =$ _____

10. $30 \div 6 =$ _____

11. $21 \div 3 =$ _____

12. $16 \div 2 =$ _____

13. $9 \div 9 =$ _____

14. $27 \div 3 =$ _____

Compare. Write >, <, or =.

15. $20 \div 5$ ___ 4

16. $36 \div 4$ ___ 7

17. $24 \div 6$ ___ 5

Problem Solving

18. A florist has 12 roses that need to be divided into 2 bouquets. If the bouquets each have the same number of roses, how many roses will be in each? _____

19. The New World Tricycle Company has received an order from one of its customers. It will take 21 wheels to build all the tricycles that the customer ordered. How many tricycles were ordered? _____

Spiral Review

20. $5,639 + 3,796 =$ _____

21. $5,639 - 3,796 =$ _____

Name_____

7·3 Relate Multiplication to Division

Write related multiplication and division sentences for each picture.

1.

2.

_____

_____

Write related multiplication and division sentences for each group of numbers.

3. 2, 4, 8

_____

4. 3, 3, 9

_____

Problem Solving

5. Janice finishes a 16-mile road race in exactly 2 hours. If she kept up a steady speed, how many miles did she run each hour?

_____

6. There are 9 cars in the parking lot, with a total of 36 wheels. If each car has the same number of wheels, how many wheels does each car have? How would you write this as a division sentence? What is a related multiplication sentence?

_____

Spiral Review

7. $4 \times 7 =$ _____

8. $8 \times 8 =$ _____

9. $9 \times 0 =$ _____

Name_____

7·4 Problem Solving: Reading for Math
Choose an Operation

Solve. Tell how you chose the operation.

1. Jacob is planting a vegetable garden. He plants 6 pepper seeds and 10 radish seeds. How many seeds does he plant in all?

   _____

2. It costs $3 to cross a toll bridge over the Warren River. If Pablo has $27, how many times can he cross the bridge?

   _____

3. Kristina notices that the kitchen floor has 9 rows of tiles and 6 columns of tiles. How many tiles in all are on the kitchen floor?

   _____

Use data from the table for problems 4–6.

Team	Wins
Jayhawks	18
Wild Wolves	16
Blue Sox	11
Pirates	7
Wildcats	2

4. How many more games did the Jayhawks win than the Blue Sox?

   _____

5. The Wild Wolves had 4 pitchers, each of whom won the same number of games. How many games did each pitcher win?   _____

6. If the Pirates had won 3 times as many games as they did, how many games would they have won?   _____

Spiral Review

7.   3
    $\times 7$

8.   9
    $\times 5$

9.   10
    $\times 6$

10.   12
    $\times 2$

Name_____

7·5 **Divide by 2**

Write a related multiplication fact. Divide.

1. 8 ÷ 2

_____

2. 14 ÷ 2

_____

3. 16 ÷ 2

_____

4. 12 ÷ 2

_____

5. 2 ÷ 2

_____

6. 2)‾6‾

_____

7. 2)‾10‾

_____

8. 2)‾4‾

_____

9. 2)‾18‾

_____

Problem Solving

10. Dennis and Vicki work together to clean out their neighbor's basement. The neighbor pays them $16. If they divide the money evenly, how much will they each get?

_____

11. Each day during the summer, Alicia goes swimming two times. So far, during the month of August, she has gone swimming 10 times. How many days have passed in August so far?

_____

12. Heather and Robbie split a small pizza that has 6 slices. If they divide the pizza equally, how many slices will each of them eat?

_____

Spiral Review

13. 7 × 4 = _____

14. 8 × 5 = _____

15. 9 + 3 = _____

Name_____

 **7·6** ▶ **Divide by 5**

Divide.

1. 20 ÷ 5 = _____   **2.** 40 ÷ 5 = _____   **3.** 30 ÷ 5 = _____

4. 10 ÷ 5 = _____   **5.** 25 ÷ 5 = _____   **6.** 15 ÷ 5 = _____

7. 5)‾4‾5‾   **8.** 5)‾2‾0‾   **9.** 5)‾3‾5‾

Problem Solving

10. Nora needs to buy 15 dinner rolls. If each package holds 5 rolls, how many packages does she need to buy?   _____

11. Jorge buys a snack at a convenience store and gets $0.35 in change. If he asks for his change in nickels, how many nickels will he get?   _____

12. A small movie theater has 70 seats total. There are 30 on the right side of the theater, and 40 on the left side. If there are 5 rows on each side of the theater, then how many seats are there in each row on the right side of the theater? How many seats are in each row on the left side?   _____

Spiral Review

13.   6,470
        − 4,567
        ‾‾‾‾‾‾‾

14.   8
          × 11
        ‾‾‾‾‾‾‾

15.   1,755
        + 1,986
        ‾‾‾‾‾‾‾

Name_____

7·7 **Problem Solving: Strategy**
Choose a Strategy

Solve.

1. For a snack, Lena has an orange that has
   12 sections in it. If she shares it equally with Cal,
   how many sections will they each have?   _____

2. The next day, Lena brings three oranges, each of
   which has 12 sections. How many sections does
   she bring in all?   _____

3. Jennifer and five friends buy 6 movie tickets for
   $8.00 each. How much do they spend in all?   _____

4. The theater sells a book of 8 tickets for $40.00.
   How much does each ticket in the book cost?   _____

Mixed Strategy Review

**Use data from the table for
problems 5–6.**

Sports Equipment Owned by Harding Elementary School	
Type of Equipment	**Number**
Kickballs	15
Softball Gloves	18
Basketballs	8
Pogo Sticks	3

5. How many more kickballs than pogo
   sticks does Harding School own?

   _____

6. If 4 students are using each basketball,
   how many students are playing basketball?   _____

Spiral Review

Compare. Write >, <, or =.

7. 7,920 ___ 8,009   8. $41.01 ___ $41.10   9. 2,767 ___ 2,767

Name_____

7·8 ▶ **Divide by 3**

Divide.

1. 12 ÷ 3 = _____   **2.** 9 ÷ 3 = _____   **3.** 18 ÷ 3 = _____

4. 27 ÷ 3 = _____   **5.** 15 ÷ 3 = _____   **6.** 12 ÷ 3 = _____

7. 3)3   **8.** 3)24   **9.** 3)6

10. 3)21   **11.** 2)18   **12.** 5)30

Problem Solving

13. There are 12 bottles of fruit juice in the refrigerator. If Teri drinks three of them every day, how many days will it take her to drink them all?   _____

14. The French flag is made up of 3 equal stripes of blue, white, and red. If a small French flag is 24 centimeters wide, then how wide will each stripe be?   _____

Spiral Review

15. A video store uses a pictograph to keep track of how many movies are rented each day of the week. The pictograph lists ◆◆◆◆◆◆next to Wednesday. If each ◆ is 10 movies, then how many movies were rented on Wednesday? _____

16. The pictograph lists ◆◆◆◆ next to Tuesday. How many movies in all were rented on Tuesday and Wednesday?   _____

Name_____

7·9 ▶ **Divide by 4**

Divide.

1. 16 ÷ 4 = _____

2. 12 ÷ 4 = _____

3. 28 ÷ 4 = _____

4. 8 ÷ 4 = _____

5. 36 ÷ 4 = _____

6. 20 ÷ 4 = _____

7. 4)‾4‾

8. 4)‾2‾4‾

9. 4)‾3‾2‾

Problem Solving

10. Behind Dean's house, there are 4 garbage cans.
    Together, the cans have 16 bags of garbage.
    If each can has the same number of bags,
    how many bags does each can have?   _____

11. Olivia is riding her bicycle around a track. It takes
    4 laps around the track to make a mile. If Olivia
    does 24 laps on her bicycle, how many miles
    does she ride?   _____

Spiral Review

12.   7
    × 6
    ____

13.   3
    × 8
    ____

14.   8
    × 3
    ____

15. 12
    × 7
    ____

16.   9
    × 3
    ____

17.   7
    × 4
    ____

18.   8
    × 5
    ____

19.   9
    × 6
    ____

 **7·10** ▶ **Divide with 0 and 1**

Divide.

1. 6 ÷ 1 = _____   **2.** 3 ÷ 3 = _____   **3.** 18 ÷ 9 = _____

4. 0 ÷ 7 = _____   **5.** 9 ÷ 1 = _____   **6.** 5 ÷ 5 = _____

7. 1)‾4‾   **8.** 6)‾2‾4‾   **9.** 8)‾0‾

10. 7)‾7‾   **11.** 1)‾8‾   **12.** 6)‾6‾

Problem Solving

13. There are 8 bicycles in the rack outside of school. At the end of the day, 8 students come outside, get on the bicycles, and ride them home. How many students are riding each bicycle?   _____

14. June, Ivan, and Tom come late to a picnic, looking for dessert. There are 0 pieces of pie left. How many pieces of pie will each of them eat?   _____

Spiral Review

15.   173
        + 92

16.   371
        − 92

17.   3
           2
        × 9

18.   9
           2
        + 3

19.   4
           3
        × 5

20.   7
           5
        + 3

21.   549
        − 75

22.   623
        + 98

Name_____

 **8·1** **Divide by 6 and 7**

Divide.

1. 6)$\overline{12}$ 2. 7)$\overline{28}$ 3. 6)$\overline{36}$

4. 6)$\overline{24}$ 5. 7)$\overline{35}$ 6. 7)$\overline{42}$

7. $48 \div 6 =$ _____   8. $42 \div 6 =$ _____   9. $21 \div 7 =$ _____

10. $27 \div 3 =$ _____   11. $14 \div 7 =$ _____   12. $30 \div 6 =$ _____

13. $6 \div 6 =$ _____   14. $54 \div 6 =$ _____   15. $35 \div 7 =$ _____

Problem Solving

16. The All-Fresh Juice Company sells its juice in packs of
    6 bottles. If Vera buys 42 bottles of All-Fresh juice, how
    many packs does she buy?   _____

17. In years other than leap years, the month of February
    has 28 days. How many weeks are in the month of
    February during these years?   _____

Spiral Review

18.   604   19.   884   20.   1,007
       − 477   + 884   − 299

Name_____

 **8·2** **Problem Solving: Reading for Math**
Solve Multistep Problems

Solve. Tell what hidden question you inferred.

1. For the first 8 days of their trip, Karen writes 4 postcards a day and Nick writes 3 postcards a day. How many postcards do they write in all?

   _____

   _____

2. For a bake sale, Anna bakes 17 muffins and Josh bakes 13 muffins. If they pack the muffins in bags of 6, how many bags can they fill?

   _____

Use data from the chart for problems 3–4.

3. Which country listed won the most medals? Which country won the fewest? What is the difference between the two totals?

   _____

4. Suppose that Sweden had won 5 times as many medals as it did, and that Jamaica had won 7 times as many medals as it did. Who would have won more? How many more?

   _____

**Medals Won at the
1996 Summer Olympics**

Country	Medals
China	50
Cuba	25
Germany	65
Jamaica	6
Sweden	8
United States	101

Spiral Review

5. 6
   × 3

6. 5
   × 7

7. 12
   + 11

8. 11
   × 11

Name_____

 **8·3** **Divide by 8 and 9**

Divide.

1. $9\overline{)45}$ 2. $8\overline{)24}$ 3. $8\overline{)48}$

4. $9\overline{)36}$ 5. $8\overline{)72}$ 6. $9\overline{)54}$

7. $7\overline{)63}$ 8. $4\overline{)32}$ 9. $9\overline{)36}$

10. $63 \div 9 =$ _____   11. $32 \div 8 =$ _____   12. $48 \div 6 =$ _____

13. $54 \div 9 =$ _____   14. $72 \div 8 =$ _____   15. $27 \div 3 =$ _____

Problem Solving

16. Pete notices that there are spiders on the steps down to the basement of his house. He counts a total of 56 legs on the spiders. Since each spider has 8 legs, how many spiders are there?   _____

17. While waiting for school to start, Tamala and Denise play some games of tic-tac-toe on a sheet of paper. For each game, they draw a frame that has 9 spaces. By the time school starts, there are 81 tic-tac-toe spaces on the sheet of paper. How many games did they play?   _____

Spiral Review

18. $9 \times 5 =$ _____   19. $35 \div 7 =$ _____   20. $8 \times 8 =$ _____

21. $21 \div 3 =$ _____   22. $6 \times 5 =$ _____   23. $16 \div 4 =$ _____

8·4 Problem Solving: Strategy
Guess and Check

Use the guess-and-check strategy to solve.

1. Each of Jane's pairs of pants has 4 pockets, and each of her jackets has 3 pockets. In packing a suitcase, she has packed clothes with a total of 17 pockets. What clothes has she packed?

_____

2. Paul has glasses that hold 5 ounces of water and cups that hold 8 ounces of water. If he pours 49 ounces of water, how many cups and glasses can he fill?

_____

3. You have 6 coins that total $0.45. What coins do you have?

_____

Mixed Strategy Review

4. Most clovers have 3 leaves, but some have 4. In a group of clovers with a total of 17 leaves, how many four-leaf clovers are there?

_____

5. There are two numbers whose difference is 6 and product is 27. What are the two numbers? _____

6. Tickets for a baseball game are $7 for lower-level seats, and $4 for upper-level seats. A ticket seller sells $25 worth of tickets. How many of each kind of ticket does she sell?

_____

Spiral Review

7. $6\overline{)24}$

8. $40 \div 5 =$ _____

 **8·5** **Explore: Dividing by 10**

Divide.

1. 10)‾20‾ **2.** 10)‾80‾ **3.** 10)‾30‾

4. 10)‾90‾ **5.** 40 ÷ 10 = _____ **6.** 70 ÷ 10 = _____

7. 10 ÷ 10 = _____ **8.** 60 ÷ 10 = _____ **9.** 50 ÷ 10 = _____

Solve.

10. After Eric finishes raking the yard, he has 20 bags full
of leaves. It takes him 10 trips to take all these bags to
the curb to be picked up. If he takes the same number
of bags on each trip, how many bags does he take
on each trip? _____

11. Rebecca buys two pens and gets $0.90 back in change.
If she asks for the change in dimes, how many dimes
will she get? _____

12. The United States Mint is issuing quarters in 50 new
designs, one for each state. If it issues the same
number of new designs every year for 10 years, how
many new designs will the Mint issue each year? _____

Spiral Review

**Find each product. Then use the Commutative Property to write a
different multiplication sentence.**

13. 4 × 7 **14.** 8 × 3 **15.** 9 × 5

_____ _____ _____

Name _____

8·6 Use a Multiplication Table to Divide

Find each missing number.
Use the multiplication table.

0	0	0	0	0	0	0	0	0	0	0	0	0
1	0	1	2	3	4	5	6	7	8	9	10	11	12
2	0	2	4	6	8	10	12	14	16	18	20	22	24
3	0	3	6	9	12	15	18	21	24	27	30	33	36
4	0	4	8	12	16	20	24	28	32	36	40	44	48
5	0	5	10	15	20	25	30	35	40	45	50	55	60
6	0	6	12	18	24	30	36	42	48	54	60	66	72
7	0	7	14	21	28	35	42	49	56	63	70	77	84
8	0	8	16	24	32	40	48	56	64	72	80	88	96
9	0	9	18	27	36	45	54	63	72	81	90	99	108
10	0	10	20	30	40	50	60	70	80	90	10	11	12
11	0	11	22	33	44	55	66	77	88	99	110	121	132
12	0	12	24	36	48	60	72	84	96	108	120	132	144
	0	1	2	3	4	5	6	7	8	9	10	11	12

1. $36 \div 4 = c$
$4 \times c = 36$

_____

2. $42 \div 6 = r$
$6 \times r = 42$

_____

3. $54 \div 9 = x$
$9 \times x = 54$

_____

4. $96 \div 8 = u$
$8 \times u = 96$

_____

Divide.

5. $24 \div 6 =$ _____

6. $35 \div 5 =$ _____

7. $88 \div 8 =$ _____

8. $8\overline{)56}$

9. $6\overline{)72}$

10. $12\overline{)60}$

Problem Solving

11. Ms. Hawthorne has 72 sheets of colored paper that she is trying to divide into 8 equal stacks. How many sheets of paper will be in each stack? _____

12. Justin measures one wall in his room and finds that it is 84 inches wide. How many feet wide is the wall? _____

Spiral Review

13. $452 - 274 =$ _____

14. $341 + 341 =$ _____

15. $807 - 237 =$ _____

16. $456 + 242 =$ _____

8·7 Use Properties and Related Facts

Write a fact family for each group of numbers.

1. 3, 5, 15

_____

_____

2. 7, 8, 56

_____

_____

3. 9, 9, 81

_____

_____

4. 7, 9, 63

_____

_____

Find each number.

5. $7 \times g = 28$ _____

6. $v \times 8 = 56$ _____

7. $9 \times w = 72$ _____

8. $45 \div b = 9$ _____

9. $d \div 6 = 4$ _____

10. $49 \div n = 7$ _____

Problem Solving

11. The aquarium has a California sea hare snail that is 30 inches long and a green turban snail that is 5 inches long. How many times longer is the sea hare than the green turban?   _____

12. Judy has 55 books that need to be placed in a bookcase with 5 shelves. If she places the same number of books on each shelf, how many books will each shelf hold?   _____

Spiral Review

Complete. Write >, <, or =.

13. 5×3 _____ 3×7 **14.** 6×8 _____ 8×6 **15.** 3×6 _____ 4×5

Name_____

9·1 Explore Multiplying Multiples of 10

Multiply.

1. $3 \times 40 =$ _____

2. $4 \times 50 =$ _____

3. $7 \times 20 =$ _____

4. $5 \times 80 =$ _____

5. $2 \times 50 =$ _____

6. $6 \times 40 =$ _____

7. $8 \times 40 =$ _____

8. $7 \times 60 =$ _____

9. $9 \times 30 =$ _____

10. $7 \times 80 =$ _____

11. $4 \times 90 =$ _____

12. $8 \times 80 =$ _____

13. $9 \times 70 =$ _____

14. $6 \times 70 =$ _____

15. $9 \times 90 =$ _____

Solve.

16. The flu has hit Traynor Elementary School. In each grade level, 20 students are out sick with the flu. If there are 6 grades in the school total, how many students are out with the flu in all?

_____

17. Scott is making flash cards. He has 50 note cards, each of which he is cutting up into 4 pieces. If he uses each piece for a flash card, how many flash cards will he have?

_____

Spiral Review

18. $35 \div 7 =$ _____

19. $8 \times 0 =$ _____

20. $9\overline{)54}$

Name_____

9·2 Multiplication Patterns

Write the number that makes each sentence true.

1. $9 \times 3 = d$ _____

$9 \times e = 270$ _____

$f \times 300 = 2,700$ _____

$9 \times 3,000 = g$ _____

2. $6 \times w = 48$ _____

$x \times 80 = 480$ _____

$6 \times 800 = y$ _____

$6 \times z = 48,000$ _____

Multiply. Use mental math.

3. $7 \times 80 =$ _____

4. $6 \times 90 =$ _____

5. $4 \times 300 =$ _____

6. $5 \times 600 =$ _____

7. $8 \times 300 =$ _____

8. $6 \times 700 =$ _____

9. $2 \times 8,000 =$ _____

10. $7 \times 4,000 =$ _____

11. $8 \times 5,000 =$ _____

Problem Solving

12. Columbus Plaza, a building in Chicago, is exactly 500 feet tall. If the owners of the building were to decide to make it 4 times taller than it currently is, how tall would it be? _____

13. Tim owns a concert hall with 4,000 seats. Next week, a singer with a new hit song is coming to town and playing 3 concerts in Tim's hall. How many tickets will Tim have for sale? _____

Spiral Review

Write related multiplication and division sentences for each group of numbers.

14. 3, 8, 24

_____

_____

15. 9, 5, 45

_____

_____

9·3 Explore Multiplying 2-Digit Numbers by 1-Digit Numbers

Write a number sentence and then solve.

1.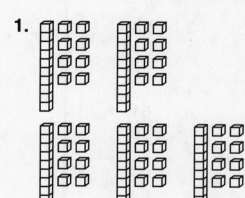

_____

2.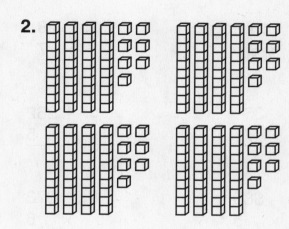

_____

Multiply.

| 3. | 16
× 4 | 4. | 21
× 4 | 5. | 28
× 6 | 6. | 25
× 7 |

Solve.

7. A bus has 14 rows of seats for passengers, with 4 seats in each row. How many passengers can be seated on the bus in all?

_____

8. There are 4 third-grade classes at Wright Elementary School. Each class has 27 students. How many third-graders are there in all?

_____

Spiral Review

| 9. | 914
− 766 | 10. | 4
3
× 6 | 11. | $84.15
+ $76.85 | 12. | 4,104
−   565 |

Name_____

Multiply 2-Digit Numbers by 1-Digit Numbers

Multiply.

1. 16
    × 5

2. 28
    × 7

3. $43
    × 4

4. 36
    × 7

5. $62
    × 5

6. 51
    × 8

7. $85
    × 5

8. $73
    × 6

9. 64
    × 9

10. $3 \times 27 =$ _____

11. $5 \times 41 =$ _____

12. $4 \times 66 =$ _____

13. $7 \times 37 =$ _____

14. $2 \times 88 =$ _____

15. $6 \times 49 =$ _____

Problem Solving

16. In the building where Elaine works, there are 6 staircases, each of which has 22 stairs in it. How many stairs are in the building in all?

_____

17. January, March, May, July, August, October, and December each have 31 days. How many days are in these months all together?

_____

Spiral Review

18. $36 \div 4 =$ _____

19. $49 \div 7 =$ _____

20. $54 \div 9 =$ _____

9·5 Estimate Products

Estimate each product.

1. 59
    × 5

2. 32
    × 7

3. 46
    × 4

4. 213
    × 7

5. 688
    × 6

6. 327
    × 8

7. 6 × 64 = _____

8. 5 × 711 = _____

9. 8 × $208 = _____

10. 9 × 291 = _____

11. 2 × $876 = _____

12. 5 × 416 = _____

13. 4 × 3,889 = _____

14. 7 × $6,119 = _____

15. 9 × 4,320 = _____

Problem Solving

16. Each of the reading textbooks for Ms. Adenauer's class has 587 pages. There are 6 reading textbooks on the shelf at the back of the classroom. About how many pages do the textbooks have in all?

    _____

17. Jamaica has an area of 4,200 square miles. The Netherlands is about 4 times bigger than Jamaica. About how big is the Netherlands?

    _____

Spiral Review

18. 1,004 − 78 = _____

19. 77 ÷ 11 = _____

20. 2 × 4 × 9 = _____

21. 428 + 36 = _____

Name_____

9·6   **Problem Solving: Reading for Math**

Find an Estimate or Exact Answer

Solve.

1. Cory delivers 44 newspapers every day. How many does he deliver in a week?

   _____

2. Elisa has promised to bring 200 crackers to school for a party. If she buys 4 packages with 58 crackers in each, will she have enough?

   _____

3. Jan buys 9 bottles of juice, each of which holds 32 ounces of juice. How many ounces of juice does she buy?

   _____

Use data from the table for problems 4–5.

Multiplex Cinema	Number of Theaters	Seats in Each Theater
Grandview	4	63
Colony	8	48
Metro	5	72

4. About how many people can attend movies at the Colony Cinema at one time?

   _____

5. Which cinema can hold more people: the Grandview or the Metro?

   _____

Spiral Review

6. 5)‾20‾   7. 32 ÷ 4 = _____   8. 6)‾42‾

Name_____

Problem Solving: Strategy
Make a Graph

Use data from the table to solve problems 1–6.

Coins in Maia's Jar	
Type of Coin	**Number**
Silver Dollars	25
Quarters	140
Dimes	200
Nickels	75
Pennies	115

1. Make a graph to show the data in the table.

2. Which type of coin does Maia have the most of? the fewest of?

_____

3. How many times more nickels than silver dollars does she have?

_____

4. What is the range in the numbers of the different types of coins? _____

5. How much money does Maia have in silver dollars and dimes? _____

6. If Maia exchanged all of her silver dollars for quarters, how would the graph change?

_____

_____

Spiral Review

7. 1,003
    − 737
   _____

8. 9
    0
   × 8
   ____

9. $34.89
   + $21.67
   _____

10. 6
     2
    × 8
   ____

Name_____

9·8 Multiply Greater Numbers

Multiply.

1.   371
     ×   3

2.   243
     ×   6

3. $5.22
     ×   4

4.   712
     ×   8

5.   645
     ×   7

6. $874
     ×   3

7. $4,792
     ×   2

8.   6,338
     ×   6

9. $49.60
     ×   8

10. $7 \times 341 =$ _____

11. $3 \times \$61.27 =$ _____

Find each product that is greater than 1,500 and less than 50,000. Use estimation to decide.

12. $9 \times 412 =$ _____

13. $6 \times 838 =$ _____

14. $7 \times 7,923 =$ _____

Problem Solving

15. The Waving Palms Hotel has 284 rooms. Each room has 4 lamps in it. How many lamps are there in all the hotel rooms combined?

_____

16. There are 3,772 insects at the Creepy-Crawly House at the City Zoo. If each insect has 6 legs, how many insect legs are there in all?

_____

Spiral Review

17. $40 \div 8 =$ _____

18. $84 \div 7 =$ _____

19. $11\overline{)121}$

10·1 Explore
Dividing Multiples of 10

Divide.

1. 3)60

2. 5)150

3. 2)140

4. 8)240

5. 7)490

6. 3)120

7. 160 ÷ 4 = _____

8. 320 ÷ 8 = _____

9. 360 ÷ 4 = _____

10. 250 ÷ 5 = _____

11. 360 ÷ 6 = _____

12. 400 ÷ 4 = _____

Solve.

13. There are 280 copies of a brand-new CD that need to be divided among 7 music stores. If each store gets the same number of copies, how many CDs will each store receive? _____

14. Jeannette and her family are taking a trip during summer vacation. They drove 480 miles in a total of 8 hours. If they kept up the same speed during the whole trip, how many miles did they drive each hour? _____

Spiral Review

Compare. Write >, <, or =.

15. 6×7 _____ 5×7

16. 4×8 _____ 8×5

17. 5×2 _____ 6×3

18. 9×8 _____ 8×9

19. 7×4 _____ 8×3

20. 9×3 _____ 7×5

Name_____

 **10·2** **Division Patterns**

Write the number that makes each sentence true.

1. $16 \div 4 = c$ _____

$160 \div 4 = d$ _____

$1{,}600 \div 4 = e$ _____

2. $27 \div 3 = k$ _____

$270 \div 3 = l$ _____

$2{,}700 \div 3\ m$ _____

Divide.

3. $4\overline{)240}$

4. $2\overline{)160}$

5. $7\overline{)210}$

6. $450 \div 9 =$ _____

7. $270 \div 3 =$ _____

8. $60 \div 3 =$ _____

Describe and complete these skip-counting patterns.

9. 180, 150, _____, 90, 60

_____

10. 50, 100, 150, _____, 250

_____

Problem Solving

11. During a project on recycling, Ms. Nahib's class brings in 270 bottles to be recycled. The bottles fill 3 recycling bins evenly. How many bottles does each bin hold?

_____

12. There are 450 tomato plants in a community garden. A total of 9 people take care of the plants, and they each take care of the same number of plants. How many tomato plants does each person take care of?

_____

Spiral Review

13. $9 \times 8 =$ _____

14. $11 \times 6 =$ _____

15. $8 \times 12 =$ _____

Name_____

 **10·3 Explore Division**

Divide.

1. 3)‾49   **2.** 5)‾62   **3.** 4)‾84

4. 71 ÷ 6 = _____   **5.** 65 ÷ 3 = _____   **6.** 47 ÷ 2 = _____

7. 72 ÷ 4 = _____   **8.** 92 ÷ 5 = _____   **9.** 86 ÷ 7 = _____

10. 121 ÷ 4 = _____   **11.** 482 ÷ 6 = _____   **12.** 720 ÷ 9 = _____

Solve.

13. A parking lot has a total of 96 spaces for cars to park.
There are 6 rows in the parking lot, each with the
same number of spaces. How many parking spaces
are in each row?   _____

14. The Jensens have a plum tree in their front yard.
They pick 84 plums from the tree and decide to
divide them evenly among 4 of their neighbors.
How many plums will each neighbor get?   _____

Spiral Review

15.   247   **16.**   37   **17.**   4,195   **18.**   195
        − 53   × 4   + 3,867   × 3

Name_____

 **10·4 Divide 2-Digit Numbers by 1-Digit Numbers**

Divide. Check your answer.

1. 5$\overline{)75}$

2. 4$\overline{)67}$

3. 2$\overline{)59}$

4. 6$\overline{)45}$

5. 4$\overline{)86}$

6. 8$\overline{)93}$

7. 88 ÷ 6 = _____

8. 49 ÷ 4 = _____

9. 51 ÷ 3 = _____

10. 74 ÷ 6 = _____

11. 64 ÷ 5 = _____

12. 58 ÷ 3 = _____

13. 80 ÷ 7 = _____

14. 95 ÷ 3 = _____

15. 67 ÷ 5 = _____

Problem Solving

16. There are 91 days in the months of April, May, and June combined. How many weeks are there in these three months?   _____

17. There are 85 ounces of juice in a bottle in the refrigerator. Darnell is giving out juice in 5-ounce glasses to his friends. How many glasses of juice will he be able to give out?   _____

Spiral Review

18. 4 × 488 = _____

19. 7 × $433 = _____

20. 8 × 621 = _____

 **10·5** **Problem Solving: Reading for Math**
Interpret the Remainder

Solve. Tell how you interpreted the remainder.

1. Antoine received $37 for his birthday and wants to spend it buying baseball cards. If each pack of cards costs $2, how many packs of cards will Antoine be able to buy?

_____

2. Dina has a set of trays that make star-shaped ice cubes. Each tray makes 6 ice cubes. She wants to have 64 ice cubes ready for a party. How many trays of ice cubes should she make?

_____

Use data from the table for problems 3–4.

3. The Bernsteins decide to divide up the carrots into equal groups to give to 4 friends. How many carrots will each friend receive? How many will be left over?

_____

Vegetables from the Bernsteins's Garden	
Vegetable	**Number**
String Beans	83
Peppers	25
Carrots	62
Radishes	32

4. What are two different ways that the Bernsteins could divide the radishes into at least 4 equal groups?

_____

Spiral Review

5. $2 \times 5 \times 8 =$ _____   **6.** $4 \times 3 \times 6 =$ _____   **7.** $2 \times 4 \times 3 \times 4 =$ _____

Name_____

 **10·6** **Estimate Quotients**

Estimate. Use compatible numbers.

1. 4)‾340‾

2. 6)‾407‾

3. 8)‾622‾

4. 2)‾535‾

5. 370 ÷ 8

6. 523 ÷ 6

7. 477 ÷ 7

8. 649 ÷ 9

9. 337 ÷ 3

Problem Solving

10. At the beach, Nick and his three sisters find 238 seashells. If they divide the shells evenly, about how many seashells will each one get?

_____

11. In Hopewell County the school year is 172 days. If each school week is 5 days long, about how many weeks are there in the school year?

_____

12. There are 612 children's books in the bookstore. If the books are divided equally among 6 bookshelves, about how many books are on each shelf?

_____

Spiral Review

13. 5 × 73 = _____

14. 6)‾54‾

15. 4 × 723 = _____

Name_____

 **10·7** **Divide 3-Digit Numbers by 1-Digit Numbers**

Divide. Check your answer.

1. 144 ÷ 3 = _____   **2.** 248 ÷ 4 = _____   **3.** 581 ÷ 7 = _____

4. 351 ÷ 4 = _____   **5.** 565 ÷ 5 = _____   **6.** 627 ÷ 8 = _____

7. 824 ÷ 5 = _____   **8.** 896 ÷ 3 = _____   **9.** 915 ÷ 4 = _____

10. 5)265   **11.** 4)384   **12.** 7)435

13. 3)650   **14.** 8)922   **15.** 4)949

Problem Solving

16. It is 195 miles from Norfolk, Virginia, to Washington, D.C. Jackie is able to drive from one city to the other in exactly 3 hours. If she keeps up a steady speed, how many miles does she travel each hour?   _____

17. There are 4 painters working together to paint the front porch of a house. When they finish the job, they receive $656 and agree to divide it evenly. How much will each painter receive?   _____

Spiral Review

18.   504
        977
      + 422
      _____

19.   9
           6
         × 2
        _____

20.   1,438
         3,779
       + 2,403
       _____

Name_____

Choose a Strategy

Solve.

1. A football field is 360 feet long from the back of one end zone to the back of the other. There are 3 feet in a yard. How many yards long is the field? _____

2. Each plastic food container can hold 5 pieces of pizza or 11 meatballs. How many pieces of pizza will fit in 8 containers? how many meatballs?

_____

3. Sports Mart has 328 baseball cards for sale, grouped 4 to a pack. How many packs does it have for sale? _____

Mixed Strategy Review

Use data from the table for problems 4–5.

4. The table above gives a list of some of the supplies that are in the school cafeteria at the end of a school week. How many eggs are left?

Food	Number per package	Number of packages
Buns	8	26
Eggs	12	11
Hot Dogs	10	15
Milk (cartons)	36	3

_____

5. The cafeteria will be serving hot dogs on Monday. Are there enough buns for all the hot dogs? _____

Spiral Review

6. 24 ÷ 3 = _____

7. 84 ÷ 7 = _____

8. 99 ÷ 11 = _____

10·9 Quotients with Zeros

Divide. Check your answer.

1. 5)52

2. 7)425

3. 2)606

4. 8)817

5. 3)611

6. 7)$7.14

7. 4)832

8. 3)926

9. 9)$909

10. 61 ÷ 2 = _____

11. 312 ÷ 3 = _____

12. 814 ÷ 4 = _____

13. 501 ÷ 5 = _____

14. $6.24 ÷ 6 = _____

15. 817 ÷ 2 = _____

Problem Solving

16. In 1933, scientists constructed a rocket that flew to a height of 3 miles. In 1955, another rocket climbed to a height of 318 miles. How many times higher did the second rocket fly?

_____

17. At a shoe factory, there are 615 shoelaces waiting to be placed into shoes. How many pairs of shoes will get shoelaces? How many shoelaces will be left over?

_____

Spiral Review

18. 6)78

19. 5 × $85 = _____

20. 99 × 9 = _____

Name_____

 **11·1** ▶ **Explore Customary Length**

Estimate and measure.

1. the width of your desk

Estimate: _____

Actual width: _____

2. the length of the blackboard

Estimate: _____

Actual length: _____

Choose the best unit of measure.

3. The sidewalk is about 25 _____ wide.
A. inches   **B.** feet   **C.** yards

4. A basketball hoop is about 3 _____ high.
A. inches   **B.** feet   **C.** yards

5. A school bus is about 40 _____ long.
A. inches   **B.** feet   **C.** yards

Solve.

6. Leslie is measuring the wall of her bedroom to see whether she can fit a bookcase and her bed next to one another. What unit and measuring tool do you think she'll use? Why?

_____

7. Andy is listening to the radio but is having a hard time hearing what's being said because of static. The announcer says that "Carter is 3 _____ taller than her younger brother." What unit did the announcer most likely say?

Spiral Review

8. $7 \times 86 =$ _____   **9.** $5\overline{)140}$ **10.** $8 \times 123 =$ _____

Name_____

 **II·2** **Customary Capacity**

Choose the best estimate.

1.

   A. 1 pt   **B.** 1 gal   **C.** 10 gal

2.

   A. 4 c   **B.** 4 pt   **C.** 4 gal

Measure the capacity of each object. Write the objects in order from least capacity to greatest.

3. soft drink can   ice chest   watering can

   _____   _____   _____

   _____

Problem Solving

4. Asa is having a party and wants to buy 12 quarts of juice. If he buys 3 gallons of juice, is he buying enough? _____

5. A 154-pound man has about 11 pints of blood. If he donates 1 pint to the Red Cross, how many *quarts* of blood will he have? _____

Spiral Review

6. $420 \div 7 =$ _____   7. $506 - 417 =$ _____   8. $83 \times 9 =$ _____

9. $330 \div 6 =$ _____   10. $626 - 529 =$ _____   11. $56 \times 7 =$ _____

II·3 Customary Weight

Choose the better estimate.

1.

A. 1 lb   **B.** 1 oz

2.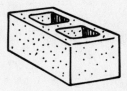

A. 20 lb   **B.** 20 oz

3.

A. 4 lb   **B.** 4 oz

4.

A. 8 lb   **B.** 8 oz

Measure the weight of each object. Write the objects in order from lightest to heaviest.

5. dinner plate   flashlight   frying pan

_____   _____   _____

_____

Problem Solving

6. Hector wants to make 10 hamburgers, each with 4 oz of hamburger meat. If he buys 2 lb of hamburger meat, will he have enough? _____

Spiral Review

7. 1,942 + 4,260 = _____

8. 4 × $744 = _____

Name_____

 **11·4 Convert Customary Units**

Complete.

1.

Pints		2		4	5
Cups	2		6		

Write the number that makes each sentence true.

2. _____ in. = 4 ft

3. 8 qt = _____ gal

4. 2 lb = _____ oz

5. _____ = 2 pt

6. 27 ft = _____ yd

7. _____ ft = 72 in.

8. _____ pt = 4 qt

9. 7 gal = _____ qt

10. 96 oz = _____ lb

Compare. Write >, <, or =.

11. 27 in. _____ 2 ft

12. 4 c _____ 1 gal

13. 3 lb _____ 48 oz

14. 4 yd _____ 10 ft

15. 32 qt _____ 9 gal

16. 7 ft _____ 84 in.

Problem Solving

17. In football, the team with the ball has to gain 10 yards in order to get a first down. On the first play that the Cougars have the ball, one of their players gains 32 feet. Did he gain enough for a first down?

_____

18. Mia buys 2 gallons of milk. How many pints of milk does she buy?

_____

Spiral Review

Compare. Write >, <, or =.

19. 5 × 9 _____ 5 × 3 × 3

20. 12 × 8 _____ 6 × 2 × 7

Name_____

II·5 ▶ Problem Solving: Reading for Math

Check for Reasonableness

Solve. Explain your answer.

1. Dan is supposed to bring 5 gallons of juice to a picnic. He shows up with 12 quarts of juice. He says that he has brought more juice than he was supposed to. Is he correct? How much juice did he bring?

   _____

2. Before a hike, Kyla and her mom weigh their packs. Kyla's weighs 21 pounds, but she does not want to carry more than 20 pounds. Her mom takes a 32-ounce pack of trail mix out of Kyla's pack and says that Kyla's pack now weighs less than 20 pounds. Is she right? How much weight in pounds did she take out of Kyla's bag?

   _____

Use data from the table for problems 3–4. Explain your answer.

3. Erin says that 4 trays of tomato plants weigh less than 1 full watering can. Is her statement reasonable?

   _____

Items Used For Gardening

Item	Weight
Brick	24 oz
Seedling	5 lbs
Tray of tomato plants	10 oz
Full watering can	3 lbs

4. Bill says that a full watering can and a brick together weigh less than a

   seedling. Is he correct? _____

Spiral Review

5. 68
   × 9

6. 8,863
   + 292

7. 2,004
   − 1,366

8. 937
   × 7

 **II·6** **Explore Metric Length**

Estimate and measure. Tell what tool you used and why.

1. the width of a piece of
notebook paper

Estimate: _____

Actual width: _____

2. the height of your desk

Estimate: _____

Actual height: _____

Choose the best estimate.

3. The ceiling is about _____ above the floor.
A. 3 cm   **B.** 3 dm   **C.** 3 m

4. A computer diskette is about _____ wide.
A. 8 cm   **B.** 8 dm   **C.** 8 m

Solve.

5. Elena is curious to find out how long her driveway is. Which metric tool and unit would she use? Why?

_____

6. Craig and Emily go to see a movie. Emily says that the the screen is 15 centimeters wide. Craig says the screen is 15 meters wide. Which measurement is more likely?

_____

Spiral Review

7. $8\overline{)49}$

8. $8 \times \$49 =$ _____

9. $455 \div 7 =$ _____

Name_____

Choose the better estimate.

1.

A. 15 mL   **B.** 15 L

2.

A. 800 mL   **B.** 800 L

Estimate. Decide if the container holds more than 1 liter, less than 1 liter, or about the same as 1 liter.

3.

_____

4.

_____

Choose the best estimate.

5. A soft drink can has about _____ of liquid.

A. 35 mL   **B.** 350 mL   **C.** 35 L

Problem Solving

6. There is a 1 L bottle of water in the refrigerator. Liana pours herself 376 mL of water. How much water is left in the bottle?

_____

7. At the store, a 2 L bottle of juice is the same price as 6 bottles that hold 325 mL of juice. Which gives you more for your money?

_____

Spiral Review

8. 5)‾732   **9.** $8,271 − $4,788 = _____   **10.** 4 × 654 = _____

Name_____

 **II·8** ► **Metric Mass**

Choose the better estimate.

1.

A. 3 g   **B.** 3 kg

2.

A. 15 g   **B.** 15 kg

Estimate. Decide if the mass of each object is greater than 1 kilogram, less than 1 kilogram, or about 1 kilogram.

3.

_____

4.

_____

Problem Solving

5. Gail's mother asks her to pick up at least 1 kg of sweet potatoes at the store. Gail picks up 6 sweet potatoes, each of which has a mass of about 200 g. Does she get enough?

_____

6. After going strawberry picking, Tamara and Doug have 4 kg of strawberries. They want to divide the berries into 8 equal portions. What will the mass of each portion be?

_____

Spiral Review

7. $780 \div 3 =$ _____

8. $6{,}045 - 2{,}445 =$ _____

9. $6\overline{)\$744}$

Name_____

 **II·9** ▶ **Convert Metric Units**

Complete.

1.

Liters	1		3	4	
Milliliters		2,000			5,000

Write the number that makes each sentence true.

2. 200 cm = _____ m

3. _____ kg = 3,000 g

4. _____ dm = 4 m

5. 5 L = _____ mL

6. _____ g = 7 kg

7. 8,000 mL = _____ L

Compare. Write >, <, or =.

8. 3,200 mL ____ 3 L

9. 479 cm ____ 5 m

10. 2,000 g ____ 2 kg

Problem Solving

11. Teresa draws three lines. The first is 13 dm long, the second is 182 cm long, and the third is 95 cm long. Which is closest to 1 m in length?

_____

12. Stan keeps his saxophone in a case. Which of the following is a likely number for the mass of the case and instrument combined: 50 g, 5 kg, or 50 kg?

_____

Spiral Review

Describe and complete the skip-counting patterns below.

13. 120, 150, _____, 210, 240

14. 420, 360, 300, _____, 180

_____   _____

 **11-10** **Problem Solving: Strategy**

Logical Reasoning

Use logical reasoning to solve.

1. There are three different-shaped objects sitting on a table in a straight line: a red triangle, a blue square, and a yellow circle. The last in line is not yellow. The first in line has four sides. What order are the objects in?

   _____

2. Nina is trying to get exactly 25 ounces of juice into a glass, but she has only a 16-ounce bottle and a 7-ounce cup to measure with. How can she use them to measure exactly 25 ounces?

   _____

   _____

   _____

   _____

3. There is a two-digit number whose product is 18 and whose sum is 9. The number in the tens place is lower than the number in the ones place. What is the two-digit number?

   _____

4. Gerardo is buying and selling baseball cards at a show. He starts out by buying 6 and then sells 3. He then buys 2 more, sells 4, and buys 3. He currently has 12 cards. How many did he start out with?

   _____

Spiral Review

Write the value of 1 in each number.

5. 213,775   6. 498,614   7. 163,823   8. 501,688

   _____   _____   _____   _____

Name_____

 **11·11** **Temperature: Fahrenheit and Celsius**

Choose the more reasonable temperature.

1. the middle of a blizzard
    A. 10°F   **B.** 10°C

2. the middle of a heat wave
    A. 37°F   **B.** 37°C

3. a pot of scalding hot water
    A. 80°F   **B.** 80°C

4. a warm spring day
    A. 73°F   **B.** 73°C

Decide which kind of clothing you should wear. Write *shirt, sweater,* or *heavy coat.*

5. 82°F

6. 14°C

7. 1°C

_____

_____

_____

8. 10°F

9. 31°C

10. 55°F

_____

_____

_____

Problem Solving

11. The highest temperature ever recorded in the United States is 57°, at Greenland Ranch, California. Is this measurement given in Celsius or Fahrenheit? _____
Source: *The Top 10 of Everything 2000*

12. On Wednesday, the temperature in Minneapolis is 73°F. By Sunday, the temperature has dropped to 39°F. What is the difference in temperature? _____

Spiral Review

13. 683 ÷ 4 = _____

14. 5)‾521‾

15. 7 × 547 = _____

Name_____

3-Dimensional Figures

Name the 3-dimensional figure the object looks like.

1.

_____

2.

_____

Copy and fold. Identify the 3-dimensional figure.

3.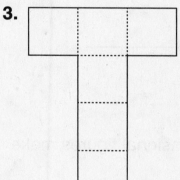

_____

4.

_____

Problem Solving

5. Francisco goes to an ice cream parlor and gets the item shown below. What 3-dimensional figures go together to make the item?

 _____

6. Which 3-dimensional figure has no faces? _____

Spiral Review

7. $8 \times 319 =$ _____   **8.** $411 \div 6 =$ _____   **9.** $5 \times \$6,378 =$ _____

 **12·2** **2-Dimensional Figures**

Complete.

	Figure	Number of Sides	Number of Angles
1.	Triangle	3	
2.		0	0
3.	Square		4

Identify each 2-dimensional figure.

4.  _____

5. _____

Problem Solving

6. The flag of Japan is shown below. What two-dimensional figures make up the flag?

 _____

7. If you cut a square in half at the points shown below, what two-dimensional shapes will you get as a result?

 _____

Spiral Review

8. 487
    × 4

9. $3,003
    − 2,426

10. 11
     2
    × 4

11. 7,275
    + 1,940

 **12·3** **Lines, Line Segments, Rays, and Angles**

Identify each figure.

1. _____

2.  _____

Decide whether the angle is less than, equal to, or greater than a right angle.

3.
_____

4.
_____

5.
_____

Problem Solving

6. Dolores gets a large pizza from Antony's Pizza. She starts on the outside of the pizza and draws a line segment across to the other side of the pizza. The line segment goes through the exact center of the pizza and is 16 inches in length. What is another word for what Dolores just measured? _____

7. Luke's bedroom is square-shaped. Each corner of his room is an angle. Are the corners of his room less than, equal to, or greater than a right angle? _____

Spiral Review

Find each missing number.

8. $6 \times 3 = t$ _____

   $6 \times u = 180$ _____

   $v \times 300 = 1,800$ _____

   $6 \times w = 18,000$ _____

9. $9 \times c = 81$ _____

   $d \times 90 = 810$ _____

   $9 \times 900 = e$ _____

   $9 \times f = 81,000$ _____

12·4 Polygons

Write *yes* or *no* to tell if each figure is a polygon. If it is, identify the polygon.

1.

_____

2.

_____

3.

_____

Problem Solving

4. The entrance to Trina's home has a rectangular door with a small triangular window on top of it. Are both the door and window polygons?

   _____

5. Ken is using toothpicks to make shapes. He places two triangles side by side as shown below, and says that he has made a hexagon. Is he correct? Why or why not?

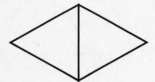 _____

_____

Spiral Review

Find the amount of change. List the coins and bills you could get.

6. Cost: $3.77. You give: $5

_____

_____

7. Cost: $2.86. You give: $10

_____

_____

Name_____

 **12·5** ➤ **Triangles**

Identify each triangle as equilateral, isosceles, or scalene.

1.
_____

2.
_____

3.
_____

4.
_____

5.
_____

6.
_____

Identify each triangle as acute, right, or obtuse.

7.
_____

8.
_____

9.
_____

Problem Solving

10. Kendra draws a triangle with sides that measure 5 inches, 6 inches, and 7 inches. What kind of triangle does she draw?   _____

11. When a square sandwich is cut into two equal pieces diagonally, two triangles are the result. Classify these triangles in 2 different ways.   _____

Spiral Review

How much time has passed?

12. Begin: 1:15 P.M.
    End: 1:50 P.M.

    _____

13. Begin: 11:16 P.M.
    End: 6:21 A.M.

    _____

14. Begin 9:37 A.M.
    End: 12:14 P.M.

    _____

Name_____

Identify each quadrilateral.

1.

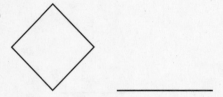 _____

2.

 _____

3.

_____

4.

 _____

5. It has 4 right angles and opposite sides that are equal.

_____

6. It has only 2 parallel sides.

_____

Problem Solving

7. Jeni makes a kite that has 4 equal sides. The kite has 2 acute angles and 2 obtuse angles. What sort of quadrilateral is the kite? _____

8. Don has 2 sticks that are each 8 inches long and 2 other sticks that are each 5 inches long. Which two quadrilaterals can he make with these sticks?

_____

Spiral Review

Write × or ÷ to make each sentence true.

9. 474 _____ 6 = 79

10. 315 _____ 4 = 1,260

11. 625 _____ 5 = 3,125

12. 392 _____ 7 = 56

12-7 **Problem Solving: Reading for Math**

Use a Diagram

Use data from the illustration to solve problems 1–3.

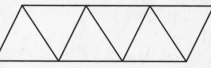

1. The illustration shows a pattern that Ms. Huerta's class wants to include in a mural on the wall of their school. How could you classify this pattern as a quadrilateral? _____

2. Each of the triangles that make up the pattern has 3 sides that are 2 feet long. How long is the top of the pattern? _____

3. Megan suggests dropping the last triangle on the right. What sort of quadrilateral would the pattern be if the class agreed with Megan's suggestion? _____

Use data from the illustration for problems 4–6.

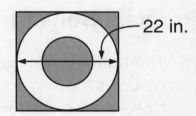

—22 in.

4. The illustration shows a pattern for a stained glass window in Gil's house. What two figures make up the illustration?

_____

5. Are any of the shaded parts of the window polygons? _____

6. The designer of the window decides to divide the window into 4 smaller squares. How long will the sides of each of these squares be? _____

Spiral Review

Compare. Write >, <, or =.

7. 49 inches ____ 4 ft   8. 222 mL ____ 2 L   9. 80 oz ____ 5 lb

Name_____

Write whether the figures are similar. Write whether the figures are congruent.

1.

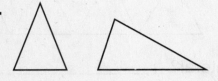

_____

2.

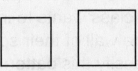

_____

3.

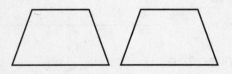

_____

4.

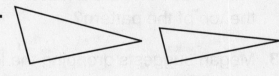

_____

Problem Solving

5. The front of a clothing store has 2 rectangular display windows. Both windows are 14 feet long and 6 feet high. Are the windows similar? Are they congruent? _____

6. Terrence is cutting triangles out of colored paper. The first triangle he cuts is an equilateral triangle with 5 inches on each side. The second triangle is an equilateral triangle with 2 inches on each side. Are the triangles similar? Are they congruent? _____

Spiral Review

7. $4\overline{)809}$

8. $5 \times 1,506 =$ _____

9. $983 \div 9 =$ _____

10. $7 \times 862 =$ _____

© McGraw-Hill School Division

Name_____

12·9 Explore Translations, Reflections, and Rotations

Write *reflection*, *rotation*, or *translation* to describe how each figure was moved.

1.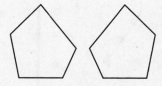

_____

2.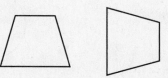

_____

3.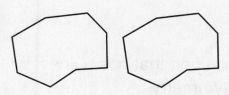

_____

4.

_____

Solve.

5. Steve is doodling on a sheet of notebook paper. He draws the pentagon on the right. He then draws it again, but so that is "pointing" right instead of up. How would you describe the relationship between the two pentagons?

_____

Spiral Review

6.
$$856 \times 9$$

7.
$$\$38.97 + \$4.16$$

8.
$$\$2,619 \times 7$$

9.
$$8,106 - 2,735$$

Name_____

Write *yes* or *no* to tell if each line is a line of symmetry.

1.

   _____

2.

   _____

3.

   _____

Draw each figure.

4. A polygon that has no lines
   of symmetry.

5. A polygon that has 3 lines
   of symmetry.

Solve.

6. There is a circular clock on the wall of Mr. Bennett's room.
   William imagines a line running through the clock from the
   12 at the top to the 6 at the bottom. Is this a line of symmetry? _____

7. LeeAnn notices a tree outside of school. The tree has
   5 branches on one side, and 4 branches on the other.
   Is the tree symmetrical? _____

Spiral Review

8. $5 \times \$3,611 =$ _____

9. $5,377 + 3,573 =$ _____

10. $8 \times 9,246 =$ _____

11. $6,436 + 2,749 =$ _____

Name_____

12·11 **Problem Solving: Strategy**
Find a Pattern

Find a pattern to solve.

1. Marisa is drawing this pattern around the edges of her notebook.

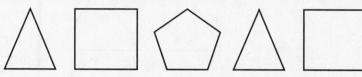

   What shape do you think she draws next? _____

2. How can you use numbers to describe the pattern that Marisa is drawing? Think of the number of sides that each polygon has.

   _____

3. Marisa continues to draw the pattern, and ends up drawing a total of 120 polygon sides. How many times does she repeat the pattern?

   _____

Mixed Strategy Review

Alfonso is using small equilateral triangles to make a large equilateral triangle. Each small triangle has sides 3 cm long. The first three rows of his triangle look like this.

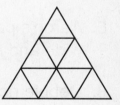

4. What is the perimeter of the figure shown? _____

5. If the pattern continues, how many small triangles would be in the sixth row? _____

Spiral Review

6. $3 \times 6 \times 2 =$ _____

7. $7 \times 3 \times 3 =$ _____

Name_____

Find each perimeter.

1.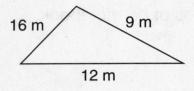
16 m   9 m
12 m

2.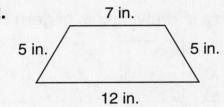
7 in.
5 in.   5 in.
12 in.

_____

3.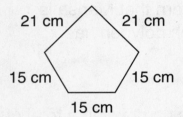
21 cm   21 cm
15 cm   15 cm
15 cm

4.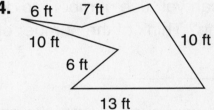
6 ft   7 ft
10 ft   10 ft
6 ft
13 ft

_____

Problem Solving

5. Aaron's parents have a dinner table that is 4 feet wide and 8 feet long. What is the perimeter of the table? _____

6. Debbie notices that the side of her house can be seen as a symmetrical pentagon. The roof is 20 feet long on both sides. The walls are 10 feet high, and the base is 24 feet wide. What is the perimeter of the side of Debbie's house? _____

Spiral Review

Compare. Write >, <, or =.

7. $61.45 ____ $60.98

8. 4 gal ____ 16 quarts

9. 3,000 dm ____ 3 m

10. 59 in. ____ 5 ft

Name_____

 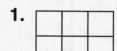 **Explore Area**

Find each area in square units.

1.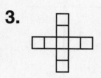

_____

2.

_____

3.

_____

4.

_____

Draw a figure with the given area. Use graph paper.

5. 15 square units

6. 28 square units

Solve.

7. Claudia draws a rectangle with 8 square units on a sheet of graph paper. She then draws a line of symmetry through the rectangle. How many square units are on each side of the line of symmetry? _____

8. Elias puts three triangles together to form a trapezoid. Each triangle has an area of 21 square units. What is the area of the trapezoid? _____

Spiral Review

Complete and describe each skip-counting pattern.

9. 490, _____, 350, 280, 210

10. 160, 240, _____, 400, 480

_____

_____

 **12·14** **Explore Volume**

Find each volume in cubic units.

1.

_____

2.

_____

Find each volume in cubic units. You may use cubes.

	Number of Rows	Number in Each Row	Number of Layers	Volume
3.	5	2	2	
4.	3	6	2	
5.	5	4	7	
6.	8	3	4	

Solve.

7. Rachel builds a shape that has 5 rows, with 4 cubes in each row. Her shape is 3 levels high. How many cubic units are in her shape?   _____

Spiral Review

Choose the better estimate. Circle your choice.

8.

A. 173 oz   **B.** 173 lb

9.

A. 27 cm   **B.** 27 m

13·1 Part of a Whole

Tell if the figure shows equal parts. If yes, write a fraction for the part that is shaded.

1.  _____

2. 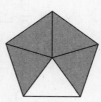 _____

Shade each figure to show the fraction.

3. $\dfrac{2}{3}$

4. $\dfrac{5}{8}$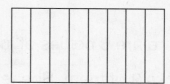

Problem Solving

5. The diagram shows how a medium pizza pie from Romano Pizzas is divided. What fraction of the pie is one slice?

_____

6. The large pizza pie from Romano Pizzas is divided into 8 slices. If Trent eats 3 slices of a large pie, what fraction of the pie does he eat? _____

Spiral Review

Name each figure.

7. _____

8. _____

Name_____

 **13·2** **Explore Equivalent Fractions**

Find an equivalent fraction. You may wish to use models.

1. $\frac{2}{8}$ _____   2. $\frac{6}{9}$ _____   3. $\frac{2}{5}$ _____

4. $\frac{3}{6}$ _____   5. $\frac{3}{4}$ _____   6. $\frac{1}{3}$ _____

7. $\frac{4}{6}$ _____   8. $\frac{9}{15}$ _____   9. $\frac{4}{14}$ _____

Solve.

10. There are 6 bottles of juice in the refrigerator. Jessie drinks 2 of them. She tells her mom that she drank $\frac{2}{6}$ of the juice. What is another way of saying this fraction?   _____

11. Eduardo is watching a wall clock. He notices that every time 5 minutes pass, the minute hand moves $\frac{1}{12}$ of the way around the clock. After 15 minutes pass, he says that $\frac{3}{12}$ of an hour has gone by. What is another way of saying this fraction?   _____

Spiral Review

12. $8\overline{)465}$

13. $6 \times \$514 =$ _____

14. $\$24.50$
    $\times \quad 4$

15. $728 \div 3 =$ _____

Name_____

 **13·3** ▶ **Compare and Order Fractions**

Compare. Write >, <, or =.

1. $\frac{1}{4}$ _____ $\frac{2}{4}$

2. $\frac{1}{2}$ _____ $\frac{4}{8}$ 3. $\frac{5}{6}$ _____ $\frac{2}{4}$ 4. $\frac{3}{4}$ _____ $\frac{7}{8}$

Tell if each fraction is closer to 0 or 1. Use the number line.

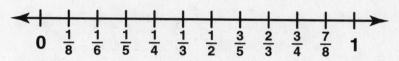

5. $\frac{1}{4}$ _____   6. $\frac{3}{4}$ _____   7. $\frac{3}{5}$ _____   8. $\frac{1}{3}$ _____

Problem Solving

9. Neil and Andrew agree to split 3 pieces of pie evenly. Neil takes 2 and gives Andrew 1. Neil says this is fair because both $\frac{2}{3}$ and $\frac{1}{3}$ are close to $\frac{1}{2}$. Is he right? Explain.

_____

10. About $\frac{1}{5}$ of the world's population lives in China. Another $\frac{1}{6}$ lives in India. Which of these fractions is closer to 1? _____

Spiral Review

11. 4 kilograms = _____ grams   12. 4 pounds = _____ ounces

Name _____

13·4 Parts of a Group

Write a fraction for the part of each group that is shaded.

1.

_____

2.

_____

3.
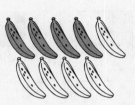

_____

Shade each set to show the fraction.

4. $\frac{2}{5}$

5. $\frac{5}{6}$

Problem Solving

6. In Ms. James's classroom there is a bag with 7 softballs. Of the softballs, 4 are white and 3 are orange. What fraction of the softballs are orange?

_____

7. Ms. James brings another white softball to school and adds it to the bag. What fraction of the softballs are orange now?

_____

Spiral Review

8. 8 quarter hours = _____ half hours

9. 180 minutes = _____ hours

10. 4 half hours = _____ hours

Name_____

 **13·5** **Explore Finding Parts of a Group**

Find each fraction of a number. You may wish to use counters.

1. $\frac{1}{2}$ of 8 _____

2. $\frac{2}{3}$ of 6 _____

3. $\frac{3}{4}$ of 12 _____

4. $\frac{4}{5}$ of 10 _____

5. $\frac{3}{8}$ of 16 _____

6. $\frac{1}{5}$ of 20 _____

7. $\frac{1}{2}$ of 18 _____

8. $\frac{3}{4}$ of 24 _____

9. $\frac{2}{3}$ of 21 _____

10. $\frac{1}{3}$ of 9 _____

11. $\frac{2}{5}$ of 25 _____

12. $\frac{5}{8}$ of 24 _____

Solve.

13. Of the 28 students in Ms. Tyson's class, $\frac{3}{4}$ have

birthdays during the school year. How many students
have birthdays during the school year?

_____

14. At a sale, Chelsea buys a CD that normally
costs $15. Because of the sale, she has to

pay only $\frac{3}{5}$ of the normal price. How much

does she pay, not counting sales tax?   _____

Spiral Review

15.   942
    ×   8

16.   8,395
    − 4,677

17. $ 6,378
    ×   6

18.   4,299
    + 3,299

Name_____

 **13·6** **Find Parts of a Group**

Find each fraction of a number. You may wish to use models.

1. $\frac{1}{2}$ of 6 _____

2. $\frac{1}{3}$ of 9 _____

3. $\frac{2}{5}$ of 10 _____

4. $\frac{3}{8}$ of 24 _____

5. $\frac{1}{5}$ of 25 _____

6. $\frac{1}{4}$ of 28 _____

7. $\frac{5}{8}$ of 16 _____

8. $\frac{3}{5}$ of 30 _____

9. $\frac{3}{4}$ of 32 _____

10. $\frac{2}{3}$ of 21 _____

11. $\frac{1}{2}$ of 18 _____

12. $\frac{4}{5}$ of 20 _____

Problem Solving

13. There are 18 olives in a jar. Denise eats $\frac{2}{3}$ of them for a snack. How

   many olives does she eat? _____

14. Roberto buys 40 sheets of notebook paper. He uses $\frac{1}{8}$ of the sheets

   doing his homework. How many sheets of notebook paper does

   Roberto use? _____

Spiral Review

Write the value of each underlined digit.

15. 4̲1,655

16. 28̲2,049

17. 81,9̲22

18. 9̲31,278

_____   _____   _____   _____

13·7 Problem Solving: Reading for Math
Check for Reasonableness

Solve. Check for reasonableness.

1. Asit cuts an apple into 10 equal pieces. He keeps $\frac{2}{5}$ and gives the rest away. How many pieces does Asit keep?

   _____

2. Of 30 cars, $\frac{1}{6}$ are blue. How many are not blue?

   _____

3. A day on the planet Neptune is about $\frac{2}{3}$ as long as a day on Earth. If an Earth day is 24 hours, how long is a day on Neptune?

   _____

Use problems 4–7 to complete the table.

4. Each team in the Park League plays 24 games. The Eagles won $\frac{5}{6}$ of their games. Fill in their record on the chart.

5. The Dragons lost $\frac{3}{4}$ of their games. Fill in their record on the chart.

Final Standings for Park League Basketball		
Team	**Wins**	**Losses**
Eagles		
Cougars		
Bruins		
Dragons		

6. The Bruins won $\frac{2}{5}$ as many games as the Eagles. Fill in their record.

7. The Cougars lost $\frac{5}{8}$ as many games as the Bruins. Fill in their record.

Spiral Review

8. $4 \times 3,855 =$ _____

9. $5,945 \div 5 =$ _____

10. $6 \times 4,625 =$ _____

11. $4,794 \div 3 =$ _____

 **13·8** **Mixed Numbers**

Write as a mixed number.

1.

_____

2.

_____

Measure to the nearest $\frac{1}{2}$ inch.

3. _____

4.

_____

Problem Solving

5. Sheila uses all of one box of spaghetti and half of another. Write a mixed number that tells how many boxes of spaghetti Sheila uses.

_____

6. To get to the beach, Rafael drives for 3 hours on the highway and $\frac{3}{4}$ of an hour on a small country road. Write a mixed number that tells how long he drives in all. _____

Spiral Review

Write *yes* or *no* to tell if each line is a line of symmetry.

7.  _____

8. _____

Name_____

13-9 ▶ Explore Adding Fractions

Add. You may wish to use fraction strips.

1. $\frac{1}{4} + \frac{2}{4} =$ _____

2. $\frac{1}{6} + \frac{3}{6} =$ _____

3. $\frac{1}{8} + \frac{3}{8} =$ _____

4. $\frac{1}{10} + \frac{3}{10} =$ _____

5. $\frac{1}{9} + \frac{4}{9} =$ _____

6. $\frac{3}{8} + \frac{3}{8} =$ _____

7. $\frac{5}{12} + \frac{1}{12} =$ _____

8. $\frac{3}{9} + \frac{5}{9} =$ _____

9. $\frac{3}{12} + \frac{4}{12} =$ _____

10. $\frac{1}{5} + \frac{2}{5} =$ _____

11. $\frac{2}{7} + \frac{3}{7} =$ _____

12. $\frac{3}{10} + \frac{5}{10} =$ _____

Solve.

13. Robert spends $\frac{1}{4}$ of an hour waiting in line at a restaurant. He then spends $\frac{2}{4}$ of an hour eating his food. How long does he spend in the restaurant in all?

_____

14. A loaf of bread is cut up into 12 equal slices. Anja and Julian each use 2 slices to make sandwiches. What fraction of the loaf do they use?

_____

Spiral Review

Compare. Write >, <, or =.

15. 3 gal _____ 12 qt

16. 6 L _____ 5,811 mL

17. 812 cm _____ 8 m

18. 49 in. _____ 4 ft

Name_____

 **13·10** **Adding Fractions**

Add. Write each sum in simplest form.

1. $\dfrac{1}{5}$
    $+\dfrac{2}{5}$
    _____

2. $\dfrac{3}{4}$
    $+\dfrac{3}{4}$
    _____

3. $\dfrac{3}{6}$
    $+\dfrac{4}{6}$
    _____

4. $\dfrac{2}{12}$
    $+\dfrac{7}{12}$
    _____

5. $\dfrac{2}{8} + \dfrac{4}{8} =$ _____

6. $\dfrac{4}{7} + \dfrac{6}{7} =$ _____

7. $\dfrac{10}{12} + \dfrac{10}{12} =$ _____

Problem Solving

8. There are two bottles of water on the kitchen counter.
   Both bottles are the same size. One is $\dfrac{3}{4}$ full and the
   other is $\dfrac{2}{4}$ full. How much water is there in all?   _____

9. Derrick drinks a glass of fruit juice that is $\dfrac{10}{12}$ full.
   He then refills the glass $\dfrac{6}{12}$ of the way with
   more juice. How much juice does he drink
   in all, measured in glasses?   _____

Spiral Review

Find the perimeter of each figure.

10.   7 in.

    4 in. [rectangle]

11.   11 cm

    9 cm [pentagon] 9 cm

    11 cm   11 cm

_____   _____

Name_____

 **13·11** ## Explore Subtracting Fractions

Subtract. You may wish to use fraction strips.

1. $\dfrac{3}{4} - \dfrac{2}{4} =$ _____

2. $\dfrac{7}{10} - \dfrac{5}{10} =$ _____

3. $\dfrac{6}{8} - \dfrac{4}{8} =$ _____

4. $\dfrac{4}{5} - \dfrac{1}{5} =$ _____

5. $\dfrac{6}{7} - \dfrac{2}{7} =$ _____

6. $\dfrac{7}{9} - \dfrac{1}{9} =$ _____

7. $\dfrac{8}{10} - \dfrac{5}{10} =$ _____

8. $\dfrac{10}{12} - \dfrac{7}{12} =$ _____

9. $\dfrac{5}{6} - \dfrac{3}{6} =$ _____

10. $\dfrac{8}{9} - \dfrac{5}{9} =$ _____

11. $\dfrac{9}{10} - \dfrac{5}{10} =$ _____

12. $\dfrac{7}{12} - \dfrac{3}{12} =$ _____

Solve.

13. A pitcher of water that is $\dfrac{7}{8}$ full is on the counter.

Amanda pours $\dfrac{1}{8}$ of the water into a glass.

How full is the pitcher now? _____

14. There is a carton of a dozen eggs that is $\dfrac{11}{12}$ full.

Matt uses $\dfrac{4}{12}$ of the carton to make pancakes.

How full is the carton now? _____

Spiral Review

15.
```
   194
   746
 + 555
```

16.
```
   9
   3
 × 3
```

17.
```
 $6,277
 ×     7
```

18.
```
  4,370
  2,139
 +  829
```

13·12 Subtracting Fractions

Subtract. Write each difference in simplest form.

1. $\dfrac{3}{8}$
    $-\dfrac{1}{8}$

2. $\dfrac{5}{6}$
    $-\dfrac{4}{6}$

3. $\dfrac{3}{4}$
    $-\dfrac{1}{4}$

4. $\dfrac{8}{10}$
    $-\dfrac{3}{10}$

5. $\dfrac{8}{9} - \dfrac{5}{9} =$

6. $\dfrac{7}{8} - \dfrac{6}{8} =$

7. $\dfrac{11}{12} - \dfrac{9}{12} =$

Problem Solving

8. Sofia's mother told her she could play outside for $\dfrac{3}{4}$ of an hour before dinner. Sofia has now played for $\dfrac{1}{4}$ of an hour. How much time does she have left to play? _____

9. The park near Adrian's house has woods covering $\dfrac{4}{8}$ of its area. The city is talking about cutting down a part of those woods equal to $\dfrac{1}{8}$ of the area of the park to make soccer fields. If that happens, how much of the park will still be covered by woods? _____

Spiral Review

Identify each figure below.

10.  _____

11.  _____

Name_____

13·13 Probability

Use the words *likely, unlikely, certain,* or *impossible*
to describe the probability.

Picking a

1. white marble _____

2. marble _____

3. bowling ball _____

4. black marble _____

Problem Solving

Use the words *likely, unlikely, certain,* or *impossible* to describe
the probability.

5. Of the 8 pieces of mail that arrived today, 1 is for Jackie. If she closes
   her eyes and picks a piece of mail, what is the probability that it is hers?

   _____

6. Luis has a game with a spinner. The spinner board is divided into 16
   equal parts. Of these, 1 part has the number 4 printed on it. The other
   parts have the numbers 1, 2, or 3 printed on them. If Luis spins the
   spinner, what is the probability that he will get a number other than 4?

   _____

Spiral Review

Round to the nearest hundred or hundred dollars.

7. $3,467 _____   8. 9,419 _____   9. 73,228 _____

13·14 Explore Finding Outcomes

List the possible outcomes. Then make the spinner and do the experiment. Record the outcomes in a line plot and a bar graph.

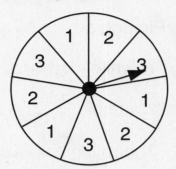

1. If you spin this spinner, what are the possible outcomes?

_____

2. Make the spinner and spin it 20 times. Record the outcomes in a line plot and a bar graph.

Solve.

3. There are 7 blackberries, 10 raspberries, and 1 gooseberry in a basket. If Lucy closes her eyes and picks a berry, which is she least likely to pick? most likely?

_____

4. Will is using the spinner shown. He needs to spin a 2 or higher to win. Is it likely that he will win?

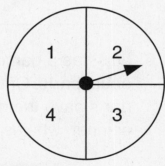

_____

Spiral Review

Identify each triangle as equilateral, isosceles, or scalene.

5.

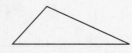

_____

6.

_____

7.

_____

13·15 Problem Solving: Strategy
Act It Out

Use data from the bar graph for problems 1–2.

1. How many times was a golf ball taken from the box?

_____

2. There are 10 golf balls left in the box. How many of each color are there? Explain your reasoning.

_____

_____

_____

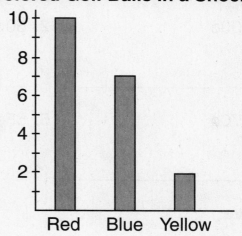

Experiment Results to Predict Colored Golf Balls in a Shoebox

Use data from the table for problems 3–4.

3. How many times was a coin picked from the piggy bank?

_____

4. Can you be certain that there are more nickels than quarters in the bank? Explain.

_____

_____

Experiment Results to Predict Coins Picked from a Piggy Bank	
Coin	**# Picked**
Penny	24
Nickel	12
Dime	3
Quarter	11

Spiral Review

5. 712 ÷ 7 = _____ **6.** 3)̄923 **7.** $816 ÷ 8 = _____

 **14·1** **Explore Fractions and Decimals**

Model each money amount as part of a dollar. Then write a fraction and decimal.

1. 30¢

2. 50¢

3. 25¢

_____   _____   _____

4. 70¢

5. 85¢

6. 20¢

_____   _____   _____

Solve.

7. Jean has 90¢. She gives away 30¢. What fraction of a dollar does she have left?

_____

8. Claire has 85¢. She spends 29¢. Write a decimal for the part of a dollar she has left.

_____

Write a decimal for the part that is shaded.

9.

10.

11.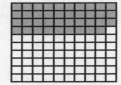

_____   _____   _____

Spiral Review

Compare. Write >, <, or =.

12. 5 lb ____ 88 oz

13. 12 ft ____ 144 in.

14. 613 cm ____ 61 m

Name_____

 14·2 **Tenths and Hundredths**

Write a decimal for each.

1. 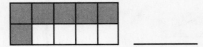 _____

2.  _____

3. $\frac{63}{100}$ _____

4. $\frac{19}{100}$ _____

5. $\frac{3}{5}$ _____

6. five tenths _____

7. thirteen hundredths _____

Write a decimal for the point. Tell if it is closest to 0, $\frac{1}{2}$, or 1.

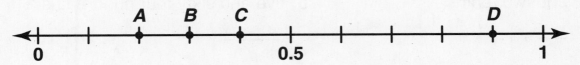

8. A _____

9. B _____

10. C _____

11. D _____

Problem Solving

12. There are 10 textbooks on the shelf in Mr. Rand's classroom. Of these, 7 are social studies textbooks. Write a fraction and decimal for the number of social studies textbooks. _____

13. There are 100 Senators in the United States Senate. In 1985, 53 of these Senators were Republicans. Write a fraction and decimal for the number of Republican Senators. _____

Spiral Review

14. $5\overline{)894}$

15. $\begin{array}{r} \$2,754 \\ \times \quad 3 \\ \hline \end{array}$

16. $\begin{array}{r} \$9,417 \\ - 5,844 \\ \hline \end{array}$

17. $\begin{array}{r} 6,294 \\ \times \quad 5 \\ \hline \end{array}$

Name_____

14·3 Decimals Greater than One

Write each decimal.

1.

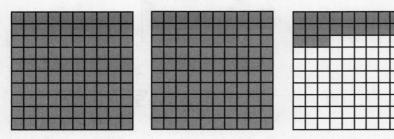

_____

2. $5\frac{4}{10}$ _____

3. $3\frac{73}{100}$ _____

4. $8\frac{25}{100}$ _____

5. two and two tenths

6. five and sixty-four hundredths

_____   _____

7.

A _____

Problem Solving

8. Julia buys a melon that weighs $3\frac{5}{10}$ pounds.

Write this number as a decimal.   _____

9. Bela keeps his baseball cards in boxes that hold
100 cards each. He has 7 boxes that are full of cards.
He has another box with 39 cards in it. Write a decimal
telling how many boxes of baseball cards Bela has.   _____

Spiral Review

Identify each triangle as acute, right, or obtuse.

10.  _____

11.  _____

Name_____

 **14·4** **Compare and Order Decimals**

Compare. Write >, <, or =.

1.   0.71 _____ 0.63

2. 2.32 _____ 2.32   **3.** 0.47 _____ 1.38   **4.** 9.13 _____ 9.08

Use the number line for problems 5–6.

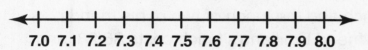

7.0 7.1 7.2 7.3 7.4 7.5 7.6 7.7 7.8 7.9 8.0

5. 7.2 _____ 7.8

6. 7.5 _____ 7.4

Problem Solving

7. A red water cooler holds 3.87 gallons of water and a blue water cooler holds 5.09 gallons of water. Which water cooler holds more?

_____

8. The population of Ireland is 3.62 million people. The population of New Zealand is 3.63 million people. The population of Lebanon is 3.51 million people. Put these countries in order of population from least to greatest.

_____

Spiral Review

Write each answer in simplest form.

9. $\dfrac{2}{9} + \dfrac{5}{9} =$ _____

10. $\dfrac{4}{7} + \dfrac{5}{7} =$ _____

11. $\dfrac{11}{12} - \dfrac{8}{12} =$ _____

Name_____

14·5 Problem Solving: Reading for Math
Choose an Operation

Solve. Tell how you chose the operation.

1. Bottles of orange juice sell for $0.89. During lunch, 9 people buy orange juice. How much do they spend on orange juice in all?

   _____

2. Mariana spent $4.17 on her lunch. Charles spent $3.49. Who spent more? How much more?

   _____

3. Abe is paying for his own lunch and the lunch of two of his friends. Abe's lunch costs $6.75, his friend Emiliano's lunch costs $6.23, and his friend Gloria's lunch costs $7.09. How much does Abe spend in all?

   _____

Use data from the table for problems 4–5.

4. Jared and Tyler are shopping for a hiking trip. Jared buys 2 pairs of socks and a cooking pot. How much does he spend? _____

Cost of Backpacking Supplies	
Item	**Cost**
Cooking Pot	$4.21
Water Bottle	$3.04
First Aid Kit	$8.96
Socks (pair)	$2.17

5. Jared has offered to buy water bottles for both of them if Tyler will buy the first aid kit. If Tyler agrees, who will spend more money? How much more?

   _____

Spiral Review

Compare. Write >, <, or =.

6. $\frac{4}{8}$ _____ $\frac{1}{2}$

7. $\frac{5}{8}$ _____ $\frac{5}{10}$

8. $\frac{4}{6}$ _____ $\frac{3}{8}$

Name_____

 **14·6 Explore Adding Decimals**

Add.

1. 0.4 + 0.5 = _____

2. 0.63 + 0.19 = _____

3. 2.44 + 0.49 = _____

4. 1.72 + 1.04 = _____

5. 2.01 + 2.47 = _____

6. 5.36 + 0.26 = _____

7. 0.75 + 0.97 = _____

8. 1.2 + 0.9 = _____

9. 1.37 + 1.37 = _____

10. 3.44 + 2.82 = _____

Solve.

11. Philip buys a grapefruit weighing 0.23 kg and a squash weighing 0.55 kg. What is the total weight of his purchases?

_____

12. The grapefruit that Philip buys costs $0.67, and the squash he buys costs $1.44. What is the total cost of his purchases?

_____

Spiral Review

Write whether the figures are similar. Write whether the figures are congruent.

13.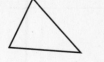

_____

14.

_____

 **14·7** **Add Decimals**

Add.

1. 3.66
   + 2.51
   _____

2. $3.96
   + 1.16
   _____

3. 9.2
   + 9.9
   _____

4. 8.92
   + 4.44
   _____

5. $3.82
   + 4.81
   _____

6. 1.83
   + 0.68
   _____

7. 0.7 + 0.4 =

_____

8. 2.5 + 5.8 =

_____

9. $4.57 + $2.24 =

_____

10. 8.22 + 8.49 =

_____

11. 0.23 + 0.68 =

_____

12. $1.78 + $0.83 =

_____

13. 6.45 + 9.64 =

_____

14. 9.65 + 0.57 =

_____

15. $6.29 + $8.88 =

_____

Problem Solving

16. Kim buys a calculator for $7.89 and a notebook
    for $0.49. How much does she spend in all?   _____

17. Brad takes a watermelon and a bag of plums to a
    picnic. The watermelon weighs 4.62 kg and the plums
    weigh 1.09 kg. How much do they weigh together?   _____

Spiral Review

18. 7 × $4,390 =

_____

19. 3,749 + 5,279 =

_____

20. 8,421 − 5,902 =

_____

Name _____

14·8 ▸ **Problem Solving: Strategy**
Solve a Simpler Problem

Solve.

1. Naomi and Ken want to rent a canoe for the afternoon. There is a $4.50 rental fee, and an additional charge of $3.25 for each hour. They rent the canoe for 3 hours. How much do they pay?

   _____

2. Brian's family is hosting a barbecue. They buy 3 bottles of apple juice for $1.75 each. They also buy 4 bottles of orange juice for $2.64 each. How much do they spend on juice in all?

   _____

Mixed Strategy Review

Use data from the table for problems 3–4.

3. The Myerson family wants to go from Grand City to Edgarton. They need to buy 2 adult tickets and 1 child's ticket. How much will they have to pay?

Bus Ticket Prices from Grand City		
Destination	**Adult**	**Child**
Brainerton	$8.75	$5.25
North Fork	$6.45	$4.30
Edgarton	$4.50	$3.10
Note: $2.00 charge for each bicycle		

   _____

4. Jenny's father is taking Jenny and her little brother to Brainerton to go bicycle riding. If they each take their bicycles with them, how much will they have to pay?   _____

Spiral Review

5. $\dfrac{5}{9} + \dfrac{7}{9} =$ _____

6. $6\overline{)904}$

7. $7 \times \$3,928 =$ _____

Name_____

 **14·9** **Explore Subtracting Decimals**

Subtract.

1. 3.77 – 1.22 = _____

2. 1.89 – 0.63 = _____

3. 0.94 – 0.56 = _____

4. 1.29 – 0.47 = _____

5. 4.25 – 0.79 = _____

6. 2.80 – 1.65 = _____

7. 8.23 – 5.81 = _____

8. 0.64 – 0.48 = _____

9. 6.29 – 4.90 = _____

10. 5.7 – 2.92 = _____

Solve.

11. Pedro goes into a store with $7.55 in his pocket. He buys a magazine for $2.97. How much money does he have left?

_____

12. It is a 1.47 km walk from Tisha's house to school. She has walked 0.39 km so far. How much farther does she have to walk?

_____

Spiral Review

Identify each figure.

13. 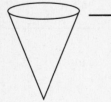 _____

14. 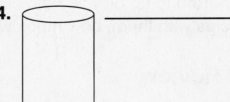 _____

Name_____

14·10 Subtract Decimals

Subtract.

1.   7.5
    − 2.2

2.   4.39
    − 2.12

3. $5.47
    − 3.65

4.   32.93
    − 4.78

5. $74.45
    − 7.90

6.   51.68
    − 1.69

7. $0.8 - 0.3 =$

_____

8. $8.7 - 4.6 =$

_____

9. $\$6.42 - \$1.98 =$

_____

10. $4.15 - 0.56 =$

_____

11. $\$9.29 - 7.64 =$

_____

12. $9.2 - 5.32 =$

_____

Problem Solving

13. At a yard sale, David notices a tape recorder for sale for $6.25. He bargains with the person selling it and gets her to take $1.40 off the price. How much does David pay for the tape recorder?

_____

14. There is a bottle containing 1.5 L of grape juice in the refrigerator. Tania pours 0.57 L into a glass for herself. How much is left in the bottle?

_____

Spiral Review

Compare. Write >, <, or =.

15. 6 qt _____ 12 pt

16. $57.23 _____ $57.30

17. $\dfrac{5}{12}$ _____ $\dfrac{1}{4}$